# Solving Equilibrium Problems with Applications to Qualitative Analysis

Steven S. Zumdahl
University of Illinois

D. C. Heath and Company
Lexington, Massachusetts Toronto

Published simultaneously in Canada.

Printed in the United States of America.

International Standard Book Number: 0-669-16718-5

10 9 8

# To the Instructor

Because the material covered in this book is treated at various levels of sophistication in general chemistry courses depending on the needs of the students, special care has been taken to make this book useful to students at all levels. The topics are organized so that the fundamental material is separated from the more advanced material. The concepts that are appropriate in a course for chemistry majors but possibly not appropriate in a service course are presented in sections that can be used at the instructor's option.

In the treatment of equilibrium, the more advanced topics have been placed in Chapter 5. The instructor can use all, none, or specific sections of Chapter 5, as is appropriate for a particular course, without effect on the material presented in Chapters 1 to 4.

In addition, equilibrium is presented in a way that allows the instructor to choose the level of mathematical sophistication appropriate for his or her students. Fundamental concepts are treated using only basic algebra. More complicated mathematical procedures are used only in the optional sections.

Chapter 6 describes the general principles of the qualitative analysis of cations and presents an exploratory experiment where the students use their own observations to design schemes of analysis for cations. The emphasis of this experiment is on creativity rather than on following recipes. A pamphlet that gives the results of all of the individual ion tests and provides suggestions on how to use this experiment can be obtained by writing the publisher.

# To the Student

Solving problems associated with chemical equilibria seems especially difficult for students. This book provides systematic, step-by-step methods for analyzing and solving these problems. Students using previous versions of this book have proven that the methods presented here really do work.

The methods are illustrated in sample problems, which are worked out for you in detail. Then there are tests to help you evaluate your understanding of the concepts. Complete solutions to the tests are given at the back of the book.

To get the maximum benefit from this book, be sure to study the introductory material and the sample problems thoroughly before attempting the test that follows that material. Be sure you do not look at the solution to a test too soon; try hard to do the test on your own. If you cannot do a test problem, review the written material and sample problems and then try the test again. After you have completed the chapter tests, do the exercises at the end of the chapter. The answers for the exercises and multiple choice questions are given in the back of the book.

Chapter 6 describes the principles and methods for the qualitative analysis of mixtures of cations. The purpose of this material is to provide a lab experience that both illustrates the scientific method and demonstrates how the principles of equilibrium can be applied to aqueous solutions. The experiment described in Chapter 6 is exploratory: It allows you to show creativity rather than to simply follow a recipe. Having to think things out on your own may be frustrating at times, but it does produce the kind of learning that will help you to better understand chemical equilibria and to become a better problem solver.

# Contents

# 1 Equilibria Involving Gaseous Reactants and Products

## CHAPTER OBJECTIVES

1. Define equilibrium, equilibrium constant, and equilibrium position.
2. Write equilibrium expressions using the Law of Mass Action.
3. Calculate the value of $K$ from a given set of equilibrium concentrations.
4. Learn to interconvert between $K$ (in units of concentration) and $K_p$ (in units of pressure).
5. Solve for equilibrium concentrations, given the initial concentrations and one equilibrium concentration.
6. Solve for equilibrium concentrations (or pressures), given the initial concentrations (or pressures) and the value of $K$.
7. Use Le Châtelier's Principle to predict shifts in equilibrium position.

## 1.1 Introduction

When substances react, the concentrations of reactants and products change continuously until the system reaches chemical equilibrium. At equilibrium, no further changes occur in the concentrations of any reactants or products as a function of time.

Equilibrium occurs because chemical reactions are reversible. To illustrate this concept, consider the very important chemical reaction for the synthesis of ammonia from dinitrogen and dihydrogen:

$$N_2(g) + 3H_2(g) \rightleftarrows 2NH_3(g) \tag{1.1}$$

The double arrows in this equation indicate not only that $N_2(g)$ and $H_2(g)$ can react to form $NH_3(g)$ but that $NH_3(g)$ can decompose to form $N_2(g)$ and $H_2(g)$. When $N_2(g)$ and $H_2(g)$ react in a closed container, the concentrations of these gases will initially decrease, and the concentration of $NH_3(g)$ will increase. As the concentrations of $N_2(g)$ and $H_2(g)$ decrease, fewer $N_2 \cdots H_2$ collisions occur, so that the forward reaction will slow down. At the same time, the increase in $NH_3$ concentration results in more $NH_3 \cdots NH_3$ collisions and the reverse reaction speeds up. Eventually the forward and reverse rates will become equal, and no further changes occur in the concentrations of either reactants or products. For example, $NH_3$ is being formed at the same rate as it is being used up, so that no change in its concentration occurs.

The following diagram shows concentration changes for the reaction just described:

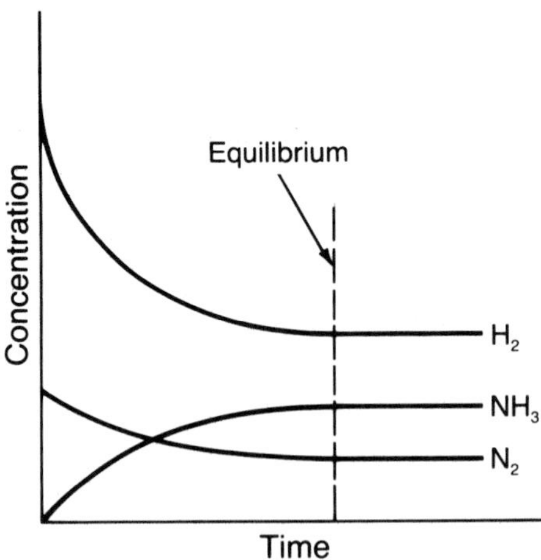


## 1.2 The Equilibrium Constant

One of the most striking properties of equilibrium is its consistency. For example, at equilibrium the concentrations of the substances in reaction (1.1) always obey the expression

$$K = \frac{[NH_3]^2}{[N_2][H_2]^3}$$

where the square brackets indicate *equilibrium* concentrations in moles per liter (mol/L). The **equilibrium constant**, $K$, is a number that is *constant* for this reaction at a particular temperature. At 500°C, its measured value is $K = 6.0 \times 10^{-2}$ $L^2/mol^2$, which means that for this reaction system at 500°C equilibrium will be achieved when the concentrations are such that

$$\frac{[NH_3]^2}{[N_2][H_2]^3} = 6.0 \times 10^{-2} \frac{L^2}{mol^2}$$

**Sample Exercise 1.1** For a particular experiment involving reaction (1.1), the equilibrium concentrations at 127°C are observed to be

$$[NH_3] = 3.05 \times 10^{-2} \text{ mol/L}$$
$$[N_2] = 8.47 \times 10^{-1} \text{ mol/L}$$
$$[H_2] = 3.05 \times 10^{-3} \text{ mol/L}$$

Calculate the value of $K$ at 127°C.

**Solution**

The equilibrium expression for this reaction is

$$K = \frac{[NH_3]^2}{[N_2][H_2]^3} = \frac{(3.05 \times 10^{-2})^2}{(8.47 \times 10^{-1})(3.05 \times 10^{-3})^3}$$
$$= 3.87 \times 10^4 \frac{L^2}{mol^2}$$

For a general reaction

$$aA + bB \rightleftarrows cC + dD$$

the equilibrium expression is written

$$K = \frac{[C]^c[D]^d}{[A]^a[B]^b}$$

where the bracketed capital letters represent concentrations of chemical species and the small letters represent the coefficients in the balanced equation. This expression is the **Law of Mass Action**, which states that the equilibrium constant expression is given by the product of the concentrations of the chemical products divided by the product of the concentrations of the reactants, each concentration being raised to the power corresponding to the coefficient of that substance in the balanced chemical equation.

**Sample Exercise 1.2**

**A.** Consider the reaction

$$2NOCl(g) \rightleftarrows 2NO(g) + Cl_2(g)$$

at 35°C, when 3.00 mol NOCl($g$), 1.00 mol NO($g$), and 2.00 mol $Cl_2$($g$) are mixed in a 10.0 L flask. After the system has reached equilibrium the concentrations are observed to be

$$[Cl_2] = 1.52 \times 10^{-1} \text{ mol/L}$$
$$[NO] = 4.00 \times 10^{-3} \text{ mol/L}$$
$$[NOCl] = 3.96 \times 10^{-1} \text{ mol/L}$$

Calculate the value of $K$ for this system at 35°C.

**Solution**

From the balanced equation for the reaction, the equilibrium expression can be written using the Law of Mass Action:

$$K = \frac{[Cl_2][NO]^2}{[NOCl]^2}$$

Substitute the measured equilibrium concentrations:

$$K = \frac{(1.52 \times 10^{-1} \text{ mol/L})(4.00 \times 10^{-3} \text{ mol/L})^2}{(3.96 \times 10^{-1} \text{ mol/L})^2}$$
$$= 1.55 \times 10^{-5} \text{ mol/L}$$

**B.** Calculate the value of $K$ for the reaction:

$$Cl_2(g) + 2NO(g) \rightleftarrows 2NOCl(g)$$

**Solution**

The equilibrium expression for this reaction is

$$K = \frac{[NOCl]^2}{[Cl_2][NO]^2}$$

$$K = \frac{(3.96 \times 10^{-1} \text{ mol/L})^2}{(1.52 \times 10^{-1} \text{ mol/L})(4.00 \times 10^{-3} \text{ mol/L})^2}$$
$$= 6.45 \times 10^{4} \text{ L/mol}$$

An easier way to solve this problem is to note that the reaction considered in this problem is the reverse of the reaction in part A, so that the equilibrium expression is the reciprocal of that in part A:

$$K = \frac{1}{1.55 \times 10^{-5}\ \text{mol/L}} = 6.45 \times 10^{4}\ \text{L/mol}$$

↖ $K$ value from part A

Sample exercise 1.2 illustrates a useful general principle: When a reaction is reversed,

$$K_{\text{reverse}} = \frac{1}{K_{\text{forward}}}$$

## The Equilibrium Constant in Terms of Pressures

Equilibria involving gases can be described in terms of either pressures or concentrations. The relationship between pressure and concentration can be seen from the ideal gas law, $PV = nRT$. Rearrangement of this equation gives

$$P = \frac{n}{V}RT = CRT$$

where $n/V$ represents the concentration of the gas in moles per liter ($C$). From this expression it can be seen that

$$P = CRT \quad \text{and} \quad C = \frac{P}{RT}$$

For the synthesis of $NH_3$,

$$K = \frac{[NH_3]^2}{[N_2][H_2]^3} = \frac{C_{NH_3}^{\,2}}{C_{N_2} \cdot C_{H_2}^{\,3}} = \frac{\left(\frac{P_{NH_3}}{RT}\right)^2}{\left(\frac{P_{N_2}}{RT}\right)\left(\frac{P_{H_2}}{RT}\right)^3} = \frac{P_{NH_3}^{\,2}}{P_{N_2} \cdot P_{H_2}^{\,3}} \cdot (RT)^2$$

$$= K_p \cdot (RT)^2$$

In this book $K$ will always represent the equilibrium constant in units of concentration, and $K_p$ will represent the equilibrium constant in units of pressure. In general, $K$ and $K_p$ will not be the same number, since the units are different. The only exception occurs when the sum of the coefficients for the products and the sum of the coefficients for the reactants are the same. In this case, the $RT$ term cancels out. For example, for the reaction

$$H_2(g) + F_2(g) \rightleftarrows 2HF(g)$$

we find that

$$K = \frac{C_{HF}^{\,2}}{C_{H_2} \cdot C_{F_2}} = \frac{\left(\frac{P_{HF}}{RT}\right)^2}{\left(\frac{P_{H_2}}{RT}\right)\left(\frac{P_{F_2}}{RT}\right)} = \frac{P_{HF}^{\,2}}{P_{H_2} \cdot P_{H_2}} = K_p$$

Note that $K$ has no units.

Substitution of $P = CRT$ into the equilibrium expression for a general reaction leads to the general relationship between $K$ and $K_p$:

$$K_p = K(RT)^{\Delta n}$$

where $\Delta n$ is the sum of the coefficients of the *gaseous* products minus the sum of the coefficients of the *gaseous* reactants.

**Sample Exercise 1.3**

Consider the reaction

$$2NOCl(g) \rightleftarrows 2NO(g) + Cl_2(g)$$

At 25°C a particular experiment showed the equilibrium pressures to be

$$P_{NOCl} = 1.2 \text{ atm}$$

$$P_{NO} = 5.0 \times 10^{-2} \text{ atm}$$

$$P_{Cl_2} = 3.0 \times 10^{-1} \text{ atm}$$

Calculate the value of $K_p$ and $K$.

**Solution**

$$K_p = \frac{P_{Cl_2} \cdot P_{NO}^2}{P_{NOCl}^2} = \frac{(3.0 \times 10^{-1} \text{ atm})(5.0 \times 10^{-2} \text{ atm})^2}{(1.2 \text{ atm})^2} = 5.2 \times 10^{-4} \text{ atm}$$

In order to use the Law of Mass Action in terms of concentration,

$$K = \frac{[Cl_2][NO]^2}{[NOCl]^2}$$

we must convert the given pressures to concentrations. Using $PV = nRT$, concentration (mol/L) $= n/V = P/RT$. For example,

$$\text{Concentration of } Cl_2 = [Cl_2] = \frac{n_{Cl_2}}{V} = \frac{P_{Cl_2}}{RT}$$

Substituting the appropriate expressions into the equilibrium constant expression,

$$K = \frac{[Cl_2][NO]^2}{[NOCl]^2} = \frac{\left(\frac{P_{Cl_2}}{RT}\right)\left(\frac{P_{NO}}{RT}\right)^2}{\left(\frac{P_{NOCl}}{RT}\right)^2} = \frac{P_{Cl_2} \cdot P_{NO}^2}{P_{NOCl}^2}\left(\frac{1}{RT}\right) = \frac{K_p}{RT}$$

$$R = 0.08206 \frac{\text{L} \cdot \text{atm}}{\text{K} \cdot \text{mol}} \qquad T = 25°\text{C} + 273 = 298 \text{ K}$$

Thus

$$K = \frac{5.2 \times 10^{-4} \text{ atm}}{RT} = \frac{5.2 \times 10^{-4} \text{ atm}}{\left(0.08206 \frac{\text{L} \cdot \text{atm}}{\text{K} \cdot \text{mol}}\right)(298 \text{ K})} = 2.1 \times 10^{-5} \frac{\text{mol}}{\text{L}}$$

Now solve this problem using the formula $K_p = K(RT)^{\Delta n}$. In this case, $\Delta n = (2 + 1) - 2 = 1$, so that

$$K_p = K(RT)$$

$$K = \frac{K_p}{RT} = 2.1 \times 10^{-5} \text{ mol/L}$$

as calculated above.

## Reactions Involving Pure Solids or Liquids

When *pure solids* or *liquids* participate in a chemical reaction, they are not included in the equilibrium expression. For example, for the reaction

$$CaCO_3(s) \rightleftarrows CaO(s) + CO_2(g)$$

the equilibrium expression is simply

$$K = [CO_2]$$

The solids, $CaCO_3$ and $CaO$, are not included.

---

**TEST 1.1**

**A.** Define equilibrium.

**B.** Write the equilibrium expression for each of the following reactions:

1. $2SO_2(g) + O_2(g) \rightleftarrows 2SO_3(g)$
2. $3O_2(g) \rightleftarrows 2O_3(g)$
3. $4NH_3(g) + 7O_2(g) \rightleftarrows 4NO_2(g) + 6H_2O(g)$
4. $Fe_2O_3(s) + 3H_2(g) \rightleftarrows 2Fe(s) + 3H_2O(g)$

**C.** Consider the hypothetical reaction

$$A(g) + 2B(g) \rightleftarrows 3C(g) + 2D(g)$$

1. Write the equilibrium expression.
2. A set of equilibrium concentrations for this system at 298 K is

$$[A] = 0.35 \text{ mol/L}$$

$$[B] = 0.20 \text{ mol/L}$$

$$[C] = 1.0 \text{ mol/L}$$

$$[D] = 0.50 \text{ mol/L}$$

Calculate $K$ and $K_p$ for this system at 298 K.

---

## 1.3 Solving for Equilibrium Concentrations

So far, the problems that have been illustrated have involved the calculation of the value of an equilibrium constant from given equilibrium concentrations. Another common type of problem involves calculation of one or more equilibrium concentrations.

These problems will be classified into two categories that require different mathematical manipulations. Your instructor will indicate which types of problems you should be able to solve.

*Type I* The given information is

1. The initial concentrations (*the concentrations before the system adjusts to come to equilibrium*), and
2. One equilibrium concentration.

These problems can be solved by using the stoichiometry of the balanced chemical equation. No algebra is required.

*Type II* The given information is

1. The initial concentrations, and
2. The value of the equilibrium constant.

The solution of these problems requires algebraic manipulations.

In solving both types of problems, these steps will be followed:

1. Write the reaction and the equilibrium constant expression.
2. Write the initial concentrations.

3. Determine the change required to reach equilibrium.
4. Calculate the equilibrium concentrations by applying the change to the initial concentrations.

## 1.4 Solving Equilibrium Problems—Type I

Principles for solving equilibrium problems will be developed in this section by considering examples where the given information consists of

1. The initial concentrations, and
2. The equilibrium concentration of one reactant or product.

Using these data, the equilibrium concentrations of the remaining reactants and products and the value of the equilibrium constant will be calculated.

Consider again the synthesis of ammonia:

$$N_2(g) + 3H_2(g) \rightleftarrows 2NH_3(g)$$

**Sample Exercise 1.4**

In a certain experiment, 1.000 mol of $N_2(g)$ and 1.000 mol of $H_2(g)$ were placed in a 1.000 L flask at 500°C and allowed to react. After the system had reached equilibrium, the flask was found to contain 0.921 mol of $N_2$. Calculate the equilibrium concentrations of $H_2$ and $NH_3$.

**Solution**

Since the equilibrium constant expression involves concentrations and not moles, first convert from moles to mol/L. This is a trivial operation in this case, since the volume is 1.000 L.

$$\text{Initial concentration of } N_2 = [N_2]_0 = \frac{1.000 \text{ mol}}{1.000 \text{ L}} = 1.000 \text{ mol/L}$$

The subscript zero means that the concentration is an initial rather than an equilibrium concentration.

For $H_2$ and $NH_3$, the initial concentrations are

$$[H_2]_0 = \frac{1.000 \text{ mol}}{1.000 \text{ L}} = 1.000 \text{ mol/L}$$

$$[NH_3]_0 = 0$$

After the system has reached equilibrium, 0.921 mol of $N_2$ remains

$$\text{Equilibrium concentration of } N_2 = [N_2] = \frac{0.921 \text{ mol}}{1.000 \text{ L}} = 0.921 \text{ mol/L}$$

The equilibrium concentrations of $H_2$ and $NH_3$ are still unknown and must be calculated. This can be done by using the stoichiometry defined by the balanced equation

$$N_2(g) + 3H_2(g) \rightleftarrows 2NH_3(g)$$

This equation shows that 1 mol of $N_2$ reacts with 3 mol of $H_2$ to produce 2 mol of $NH_3$.

In this problem, how much $N_2$ has been consumed? The amount of $N_2$ changes from 1.000 mol to 0.921 mol as the system proceeds to equilibrium. Thus 1.000 − 0.921 = 0.079 mol of $N_2$ has reacted. The number of moles of $H_2$ consumed can be calculated using normal stoichiometric procedures:

$$0.079 \text{ mol } N_2 \times \frac{3 \text{ mol } H_2}{1 \text{ mol } N_2} = 0.237 \text{ mol } H_2 \text{ consumed}$$

Similarly, the amount of $NH_3$ produced can be calculated:

$$0.079 \text{ mol } N_2 \times \frac{2 \text{ mol } NH_3}{1 \text{ mol } N_2} = 0.158 \text{ mol } NH_3 \text{ produced}$$

What are the equilibrium concentrations of $NH_3$ and $H_2$? The flask originally contained no $NH_3$, so the amount of $NH_3$ present at equilibrium is 0.158 mol. Since the volume is 1.000 L, at equilibrium,

$$[NH_3] = [NH_3]_0 \text{ plus } \left(\begin{array}{l}\text{the change needed to} \\ \text{reach equilibrium}\end{array}\right) = 0 + 0.158 \text{ mol/L}$$

The original 1.000 mol of hydrogen present has been reduced by 0.237 mol; at equilibrium

$$[H_2] = [H_2]_0 \text{ plus } \left(\begin{array}{l}\text{the change to reach} \\ \text{equilibrium}\end{array}\right) = 1.000 \text{ mol/L} - 0.237 \text{ mol/L}$$

$$= 0.763 \text{ mol/L}$$

Note that the change in $H_2$ is negative since $H_2$ is consumed to reach equilibrium.

**TEST 1.2** In a certain experiment, 1.000 mol of $NH_3$ was placed in an empty 1.000 L flask at 500°C. After equilibrium was reached, 0.399 mol of $N_2$ was found to be in the flask. Calculate the concentrations of $NH_3$ and $H_2$ at equilibrium.

**Sample Exercise 1.5** In an experiment, 2.00 mol of $N_2$, 1.00 mol of $H_2$ and 3.00 mol of $NH_3$ were mixed in a 1.00 L flask at 500°C. At equilibrium, 2.77 mol of $H_2$ were found to be present in the flask. Calculate the concentrations of $N_2$ and $NH_3$ at equilibrium.

**Solution**

In this experiment, the amount of $H_2$ changes from 1.00 mol to 2.77 mol. Thus $2.77 - 1.00 = 1.77$ mol of $H_2$ has been formed as the system proceeds to equilibrium. For this to occur, $NH_3$ must have been consumed to produce $N_2$ and $H_2$. The number of moles of $N_2$ produced can be calculated:

$$1.77 \text{ mol } H_2 \times \frac{1 \text{ mol } N_2}{3 \text{ mol } H_2} = 0.590 \text{ mol } N_2 \text{ produced}$$

Since the volume is 1.00 L, 0.590 mol/L of $N_2$ was formed to reach equilibrium. The concentration of $N_2$ at equilibrium is given by the initial concentration plus the change to reach equilibrium:

$$[N_2] = [N_2]_0 + \text{change} = 2.00 \text{ mol/L} + 0.590 \text{ mol/L}$$

$$= 2.59 \text{ mol/L}$$

The number of moles of $NH_3$ consumed is given by

$$1.77 \text{ mol } H_2 \times \frac{2 \text{ mol } NH_3}{3 \text{ mol } H_2} = 1.18 \text{ mol } NH_3 \text{ consumed}$$

Since the volume is 1.00 L, 1.18 mol/L of $NH_3$ are consumed. At equilibrium,

$$[NH_3] = [NH_3]_0 \text{ plus change} = 3.00 \text{ mol/L} + 1.18 \text{ mol/L}$$
$$= 1.82 \text{ mol/L}$$

### Summary

In doing problems of the type considered above, it is best to proceed as follows:

1. Write the balanced equation for the reaction.
2. Write the initial concentrations.
3. Calculate the change required to reach equilibrium.
4. Apply the change to the initial concentrations to obtain the equilibrium concentrations.

### Calculations Involving Pressures

Note that equilibria involving pressures rather than concentrations can be solved using the procedures introduced above, since at constant $T$, pressure is proportional to concentration for an ideal gas.

**TEST 1.3** A quantity of $N_2O_4(g)$ is introduced into a flask at an initial pressure of 2.0 atm at temperature $T$. After the $N_2O_4(g)$ has decomposed to $NO_2(g)$ and has come to equilibrium, the pressure of $N_2O_4(g)$ is 1.8 atm. Calculate the value of $K_p$ for the process

$$N_2O_4(g) \rightleftarrows 2NO_2(g)$$

at temperature $T$.

**A.** Set up the expression for $K_p$ in terms of the pressures of $N_2O_4(g)$ and $NO_2(g)$.
**B.** Calculate the pressure of $N_2O_4(g)$ consumed when the reaction comes to equilibrium and the equilibrium pressure of $N_2O_4(g)$.
**C.** What is the pressure of $NO_2(g)$ formed to reach equilibrium?
**D.** Calculate $K_p$ using the equilibrium pressures of $NO_2(g)$ and $N_2O_4(g)$.

## 1.5 The Equilibrium Position

It is important to recognize that although the special *ratio* of the concentrations of products to reactants defined by the equilibrium constant expression is constant, the equilibrium concentrations will not always be the same. To illustrate this point, consider the results of the calculations carried out in sample exercises 1.4 and 1.5 and in test 1.2. These are summarized in Table 1.1.

**TEST 1.4** Show that the ratio $[NH_3]^2/[H_2]^3[N_2]$ is constant at equilibrium for each of the situations shown in Table 1.1.

**Table 1.1** Initial and equilibrium concentrations in the ammonia synthesis reaction

| Experiment | Initial Concentrations | Equilibrium Concentrations |
|---|---|---|
| I | $[N_2]_0 = 1.000$ mol/L<br>$[H_2]_0 = 1.000$ mol/L<br>$[NH_3]_0 = 0$ | $[N_2] = 0.921$ mol/L<br>$[H_2] = 0.763$ mol/L<br>$[NH_3] = 0.158$ mol/L |
| II | $[N_2]_0 = 0$<br>$[H_2]_0 = 0$<br>$[NH_3]_0 = 1.000$ mol/L | $[N_2] = 0.399$ mol/L<br>$[H_2] = 1.197$ mol/L<br>$[NH_3] = 0.202$ mol/L |
| III | $[N_2]_0 = 2.00$ mol/L<br>$[H_2]_0 = 1.00$ mol/L<br>$[NH_3]_0 = 3.00$ mol/L | $[N_2] = 2.59$ mol/L<br>$[H_2] = 2.77$ mol/L<br>$[NH_3] = 1.82$ mol/L |

From the results of test 1.4, note that the ratio $[NH_3]^2/[N_2][H_2]^3$ is constant within errors caused by round-off for these experiments, even though the equilibrium concentrations are quite different for different experiments.

Each set of equilibrium concentrations is called an **equilibrium position**. It is very important to distinguish between an equilibrium position and the equilibrium constant. The equilibrium position depends on the initial concentrations, but the equilibrium constant is always the same at a given temperature. There are an infinite number of equilibrium positions, depending on the initial conditions, but only one equilibrium constant for a given reaction system at a given temperature.

**TEST 1.5** Consider the reaction

$$2SO_2(g) + O_2(g) \rightleftarrows 2SO_3(g)$$

at 600°C.

**A.** In a certain experiment, 2.00 mol of $SO_2$, 1.50 mol of $O_2$, and 3.00 mol of $SO_3$ were placed in a 1.00 L vessel. After the system reached equilibrium, 3.50 mol of $SO_3$ were found to be present. Calculate
1. The equilibrium concentrations of $O_2$ and $SO_2$.
2. $K$.
3. $K_p$.

**B.** In a different experiment at the same temperature (600°C), 0.500 mol of $SO_2$ and 0.350 mol of $SO_3$ were mixed in a 1.000 L container. When the system reached equilibrium, 0.045 mol of $O_2$ was found to be present. Calculate
1. The equilibrium concentrations of $SO_2$ and $SO_3$.
2. $K$.

## 1.6 The Reaction Quotient

The strategy that has been developed for solving equilibrium problems involves writing the initial concentrations and then defining the change in concentration needed to reach equilibrium. To do this, the direction of the shift from initial to equilibrium concentrations must be known.

Determination of the direction of the shift is easy when the initial concentration of a reactant or a product is zero. On the other hand, in problems where all initial concentrations are nonzero, the direction of the adjustment to equilibrium may not be obvious.

To determine the direction of the shift in such cases, the **reaction quotient**, $Q$, will be used. The reaction quotient has the same form as the equilibrium constant expression except that *initial* concentrations are used instead of equilibrium concentrations. For the synthesis of ammonia, $N_2(g) + 3H_2(g) \rightleftarrows 2NH_3(g)$,

$$Q = \frac{[NH_3]_0^{\,2}}{[N_2]_0[H_2]_0^{\,3}}$$

To determine in which direction the system will proceed from the initial concentrations toward equilibrium, one needs to compare $Q$ and $K$.

1. $Q$ is greater than $K$.

   If $Q$ is greater than $K$, the ratio of initial concentrations of products to reactants is too large. To reach equilibrium, products must be changed to reactants. The system proceeds to the left (consuming products, forming reactants) to reach equilibrium. In this case, the system shifts to the left.
2. $Q$ is less than $K$.

   If $Q$ is less than $K$, the system must proceed in the direction of more products to reach equilibrium. In this case, the system shifts to the right.
3. $Q$ is equal to $K$.

   The initial concentrations are equilibrium concentrations. No shift will occur.

**Sample Exercise 1.6** For the synthesis of ammonia, the value of $K$ is $6 \times 10^{-2}\ L^2/mol^2$ at 500°C. In an experiment, 0.50 mol of $N_2(g)$, $1.0 \times 10^{-2}$ mol of $H_2(g)$, and $1.0 \times 10^{-4}$ mol of $NH_3(g)$ are mixed at 500°C in a 1.0 L flask. In which direction will the system proceed to reach equilibrium?

**Solution**

The initial concentrations are

$$[NH_3]_0 = 1.0 \times 10^{-4}\ mol/L$$
$$[N_2]_0 = 5.0 \times 10^{-1}\ mol/L$$
$$[H_2]_0 = 1.0 \times 10^{-2}\ mol/L$$

Calculate the reaction quotient:

$$Q = \frac{[NH_3]_0^{\,2}}{[N_2]_0[H_2]_0^{\,3}} = \frac{(1.0 \times 10^{-4})^2}{(5.0 \times 10^{-1})(1.0 \times 10^{-2})^3} = 2.0 \times 10^{-2}\ L^2/mol^2$$

$$K = 6 \times 10^{-2}\ L^2/mol^2$$

Thus $Q$ is less than $K$ and the system will proceed to the right ($N_2$ and $H_2$ consumed, $NH_3$ formed) to reach equilibrium.

**TEST 1.6** Consider the reaction

$$2NOCl(g) \rightleftarrows 2NO(g) + Cl_2(g)$$

where $K = 1.55 \times 10^{-5}$ mol/L at 35°C. In an experiment, $1.0 \times 10^{-1}$ mol of $NOCl(g)$, $1.0 \times 10^{-3}$ mol of $NO(g)$, and $1.0 \times 10^{-4}$ mol of $Cl_2(g)$ are mixed at 35°C in a 2.0 L flask. In which direction will the system proceed to reach equilibrium?

## 1.7 Solving Equilibrium Problems—Type II

In this section, techniques will be developed for solving problems where the given information is

1. The initial concentrations, and
2. The value of $K$.

**Sample Exercise 1.7**

At 700 K, carbon monoxide reacts with water to form carbon dioxide and hydrogen:

$$CO(g) + H_2O(g) \rightleftarrows CO_2(g) + H_2(g)$$

The equilibrium constant for this reaction at 700 K is 5.10.

Consider an experiment in which 1.00 mol of $CO(g)$ and 1.00 mol of $H_2O(g)$ are mixed together in a 1.00 L flask at 700 K. Calculate the concentrations of all species at equilibrium.

**Solution**

The first step in solving equilibrium problems is to write the equilibrium expression. In this case,

$$K = \frac{[CO_2][H_2]}{[CO][H_2O]} = 5.10$$

Next write down the initial concentrations:

$$[CO]_0 = \frac{1.00 \text{ mol}}{1.00 \text{ L}} = 1.00 \text{ mol/L}$$

$$[H_2O]_0 = \frac{1.00 \text{ mol}}{1.00 \text{ L}} = 1.00 \text{ mol/L}$$

$$[CO_2]_0 = 0$$

$$[H_2]_0 = 0$$

It is clear that for this system to reach equilibrium it must proceed to the right (some $CO_2$ and $H_2$ will be formed by consuming some CO and $H_2O$). The key question is: How much $CO_2$ and $H_2$ will form as the system comes to equilibrium?

To consider this question systematically, use algebraic techniques. As is customary in algebra, assign the symbol $x$ to the unknown quantity. In this problem, the amounts of $CO_2$ and $H_2$ formed to reach equilibrium are the unknown quantities. Let $x$ equal the number of moles per liter of $CO_2$ formed to reach equilibrium.

If $x$ mol/L of $CO_2$ have been formed, how much $H_2$ will be present? Since $CO_2$ and $H_2$ are formed in equal amounts, $x$ mol/L of $H_2$ will be formed.

How much CO is consumed to form $x$ mol/L of $CO_2$ and $H_2$? Since all the coefficients in the balanced equation are 1, $x$ mol/L of CO are consumed, as well as $x$ mol/L of $H_2O$. Thus, at equilibrium, the concentrations of reactants and products present are

Concentration of substance at equilibrium = concentration initially present + change

$$[CO] = [CO]_0 - x = 1.00 \text{ mol/L} - x \text{ mol/L}$$

$$[H_2O] = [H_2O]_0 - x = 1.00 \text{ mol/L} - x \text{ mol/L}$$

$$[CO_2] = [CO_2]_0 + x = 0 + x \text{ mol/L}$$

$$[H_2] = [H_2]_0 + x = 0 + x \text{ mol/L}$$

The ratio of these equilibrium concentrations must give the equilibrium constant:

$$K = 5.10 = \frac{[CO_2][H_2]}{[CO][H_2O]} = \frac{(x)(x)}{(1.00 - x)(1.00 - x)} = \frac{x^2}{(1.00 - x)^2}$$

Taking the square root of both sides gives

$$\frac{x}{1 - x} = \sqrt{5.10} = 2.26$$

$$x = 2.26 - 2.26x$$

$$3.26x = 2.26$$

$$x = 0.69 \text{ mol/L}$$

Thus the equilibrium concentrations are

$$[CO_2] = x = 0.69 \text{ mol/L}$$

$$[H_2] = x = 0.69 \text{ mol/L}$$

$$[H_2O] = 1.00 - x = 0.31 \text{ mol/L}$$

$$[CO] = 1.00 - x = 0.31 \text{ mol/L}$$

**TEST 1.7** Consider the reaction at 700 K:

$$CO(g) + H_2O(g) \rightleftarrows CO_2(g) + H_2(g) \qquad K = 5.10$$

Calculate the equilibrium concentrations of all reactants and products in an experiment where 3.00 mol of $CO(g)$ and 3.00 mol of $H_2O(g)$ are mixed in a 1.00 L flask at 700 K.

**Sample Exercise 1.8** Consider the reaction

$$H_2(g) + F_2(g) \rightleftarrows 2HF(g)$$

at a temperature where $K = 1.15 \times 10^2$. In an experiment, 3.00 mol of $H_2$ and 6.00 mol of $F_2$ are mixed in a 3.00 L container. Calculate the equilibrium concentrations of $H_2$, $F_2$, and HF.

**Solution**

The equilibrium constant expression is

$$K = 1.15 \times 10^2 = \frac{[HF]^2}{[H_2][F_2]}$$

The initial concentrations are

$$[H_2]_0 = \frac{3.00 \text{ mol}}{3.00 \text{ L}} = 1.00 \text{ mol/L}$$

$$[F_2]_0 = \frac{6.00 \text{ mol}}{3.00 \text{ L}} = 2.00 \text{ mol/L}$$

$$[HF]_0 = 0$$

Now define the change required to reach equilibrium. In this case, since no HF is initially present, some $H_2$ and $F_2$ will react to form HF.

Let $x$ be the concentration of $H_2$ consumed. Using the stoichiometry of the reaction,

$$H_2 \quad + \quad F_2 \quad \longrightarrow \quad 2HF$$

$$x \text{ mol/L } H_2 + x \text{ mol/L } F_2 \longrightarrow 2x \text{ mol/L HF}$$

These are the changes in concentration required to reach equilibrium.

Now write the equilibrium concentrations:

$$[H_2] = [H_2]_0 + \text{change} = 1.00 \text{ mol/L} - x \text{ mol/L}$$

$$[F_2] = [F_2]_0 + \text{change} = 2.00 \text{ mol/L} - x \text{ mol/L}$$

$$[HF] = [HF]_0 + \text{change} = 0 + 2x \text{ mol/L}$$

To solve for the value of $x$, substitute these concentrations in the equilibrium expression:

$$K = 1.15 \times 10^2 = \frac{[HF]^2}{[H_2][F_2]} = \frac{(2x)^2}{(1.00 - x)(2.00 - x)}$$

Note that this expression cannot be solved by taking the square root of both sides, since the denominator is not a perfect square. To solve for $x$, first multiply out the terms:

$$1.15 \times 10^2 = \frac{(2x)^2}{(1.00 - x)(2.00 - x)} = \frac{4x^2}{2.00 - 3.00x + x^2}$$

$$(1.15 \times 10^2)(2.00 - 3.00x + x^2) = 4x^2$$

$$(2.30 \times 10^2) - (3.45 \times 10^2)x + (1.15 \times 10^2)x^2 = 4x^2$$

Subtracting $4x^2$ from both sides and rearranging gives

$$(1.11 \times 10^2)x^2 - (3.45 \times 10^2)x + (2.30 \times 10^2) = 0$$

This is a quadratic equation, which has the general form

$$ax^2 + bx + c = 0$$

The solutions to an equation of this type can be obtained from the formula

$$x = \frac{-b \pm \sqrt{b^2 - 4ac}}{2a}$$

where $a$ and $b$ are the coefficients of $x^2$ and $x$, respectively, and $c$ is the constant in the general equation. In the quadratic equation above,

$$a = 1.11 \times 10^2$$

$$b = -3.45 \times 10^2$$

$$c = 2.30 \times 10^2$$

Thus

$$x = \frac{-(-3.45 \times 10^2) \pm \sqrt{(-3.45 \times 10^2)^2 - (4)(1.11 \times 10^2)(2.30 \times 10^2)}}{2(1.11 \times 10^2)}$$

$$= \frac{3.45 \times 10^2 \pm \sqrt{1.19 \times 10^5 - 1.02 \times 10^5}}{2.22 \times 10^2}$$

$$= \frac{3.45 \times 10^2 \pm \sqrt{1.7 \times 10^4}}{2.22 \times 10^2} = \frac{3.45 \times 10^2 \pm 1.30 \times 10^2}{2.22 \times 10^2}$$

The two roots are

$$x = \frac{3.45 \times 10^2 + 1.30 \times 10^2}{2.22 \times 10^2} = 2.14 \text{ mol/L}$$

$$x = \frac{3.45 \times 10^2 - 1.30 \times 10^2}{2.22 \times 10^2} = 0.968 \text{ mol/L}$$

In this situation, there is only one correct value of $x$ (there is only one equilibrium position which results from these initial concentrations). Which value of $x$ is correct?

The correct value can be chosen by looking at the equilibrium concentrations. Since

$$[H_2] = 1.00 - x$$

the value $x = 2.14$ gives a negative number for the concentration of $H_2$. This is physically impossible. Therefore, $x = 0.968$ mol/L must be the correct root.

Now compute the equilibrium concentrations:

$$[H_2] = 1.00 - x = 1.00 - 0.968 = 3.2 \times 10^{-2} \text{ mol/L}$$

$$[F_2] = 2.00 - x = 2.00 - 0.968 = 1.03 \text{ mol/L}$$

$$[HF] = 2x = 2(0.968) = 1.94 \text{ mol/L}$$

After solving the quadratic equation, it is wise to check the correctness of the computed concentrations. This can be done by substituting into the equilibrium constant expression:

$$\frac{[HF]^2}{[H_2][F_2]} = \frac{(1.94)^2}{(3.2 \times 10^{-2})(1.03)} = 114$$

The given value of $K$ is 115. Thus there is agreement within the error caused by round-off, and the calculated concentrations are shown to be correct.

**TEST 1.8** Consider the reaction

$$H_2(g) + I_2(g) \rightleftarrows 2HI(g)$$

at a temperature where $K = 60.0$. In an experiment, 1.50 mol of $H_2$ and 2.50 mol of $I_2$ are placed in a 1.00 L flask. Calculate the equilibrium concentrations of $H_2$, $I_2$, and HI.

Next, problems in which none of the initial concentrations are zero will be considered. To solve these problems,

1. Calculate the reaction quotient, $Q$.
2. Determine the direction of the shift to equilibrium.
3. Define the change required to reach equilibrium.
4. Calculate the equilibrium concentrations.

**Sample Exercise 1.9** Consider the reaction

$$2HF(g) \rightleftarrows H_2(g) + F_2(g)$$

where $K = 1.0 \times 10^{-2}$ at some very high temperature. In an experiment, 5.00 mol of HF(*g*), 0.500 mol of $H_2(g)$, and 0.750 mol of $F_2(g)$ are mixed in a 5.00 L flask and allowed to react to equilibrium.

**A.** Write the equilibrium constant expression.

**Solution**

$$K = \frac{[H_2][F_2]}{[HF]^2} = 1.0 \times 10^{-2}$$

**B.** Calculate the initial concentrations.

**Solution**

$$[HF]_0 = \frac{5.00 \text{ mol}}{5.00 \text{ L}} = 1.00 \text{ mol/L}$$

$$[F_2]_0 = \frac{0.750 \text{ mol}}{5.00 \text{ L}} = 0.150 \text{ mol/L}$$

$$[H_2]_0 = \frac{0.500 \text{ mol}}{5.00 \text{ L}} = 0.100 \text{ mol/L}$$

**C.** Calculate $Q$ and determine in which direction the reaction will shift to reach equilibrium.

**Solution**

$$Q = \frac{[F_2]_0[H_2]_0}{[HF]_0^2} = \frac{(0.150)(0.100)}{(1.00)^2} = 1.50 \times 10^{-2}$$

$Q$ is larger than $K$. To come to equilibrium, the system must shift to the left (i.e., the numerator must get smaller and the denominator larger).

**D.** Define the change required to reach equilibrium.

**Solution**

Let $x$ be the mol/L of $H_2$ consumed to reach equilibrium. Note that $x$ mol/L of $F_2$ will also be consumed and that $2x$ mol/L of HF will be formed.

**E.** Define the equilibrium concentrations of all species.

**Solution**

$$[H_2] = [H_2]_0 - x = 0.100 - x$$

$$[F_2] = [F_2]_0 - x = 0.150 - x$$

$$[HF] = [HF]_0 + 2x = 1.00 + 2x$$

**F.** Solve for $x$ and calculate the equilibrium concentrations.

**Solution**

$$1.00 \times 10^{-2} = \frac{[H_2][F_2]}{[HF]^2} = \frac{(0.100 - x)(0.150 - x)}{(1.00 + 2x)^2}$$

Multiplying and collecting terms gives

$$0.960x^2 - 0.290x + 5.00 \times 10^{-3} = 0$$

This is a quadratic equation where

$$a = 0.960$$

$$b = -0.290$$

$$c = 5.00 \times 10^{-3}$$

Use the quadratic formula to solve for $x$.

$$x = \frac{-b \pm \sqrt{b^2 - 4ac}}{2a} = \frac{-(-0.290) \pm \sqrt{(-0.290)^2 - 4(0.960)(5.00 \times 10^{-3})}}{2(0.960)}$$

Solving for $x$ gives the roots

$$x = 0.284 \text{ mol/L}$$

$$x = 0.018 \text{ mol/L}$$

Since $[H_2] = 0.100 - x$, the root $x = 0.284$ cannot be correct. Thus

$$x = 0.018 \text{ mol/L}$$

is the appropriate root. The equilibrium concentrations are

$$[H_2] = [H_2]_0 - x = 0.100 \text{ mol/L} - 0.018 \text{ mol/L} = 0.082 \text{ mol/L}$$

$$[F_2] = [F_2]_0 - x = 0.150 \text{ mol/L} - 0.018 \text{ mol/L} = 0.132 \text{ mol/L}$$

$$[HF] = [HF]_0 + 2x = 1.00 \text{ mol/L} + 2(0.018 \text{ mol/L}) = 1.04 \text{ mol/L}$$

**TEST 1.9** Consider the reaction

$$PCl_5(g) \rightleftarrows PCl_3(g) + Cl_2(g) \qquad K = 5.0 \times 10^{-2} \text{ mol/L at } 240°C.$$

In an experiment, 0.200 mol of $PCl_5(g)$, 0.500 mol of $PCl_3(g)$, and 0.300 mol of $Cl_2(g)$ are mixed in a 2.00 L vessel.

**A.** Calculate the initial concentrations.
**B.** Calculate $Q$ using these initial concentrations.
**C.** Which way does the system adjust to reach equilibrium?
**D.** Define the change needed to reach equilibrium in terms of $x$.
**E.** Define the equilibrium concentrations in terms of $x$.
**F.** Substitute the equilibrium values into the equilibrium expression and solve for $x$.
**G.** Calculate the equilibrium concentrations of $PCl_5(g)$, $PCl_3(g)$, and $Cl_2(g)$.

## Calculations Involving Pressures

**Sample Exercise 1.10** Consider the reaction

$$N_2O_4(g) \rightleftarrows 2NO_2(g)$$

at a temperature where $K_p = 0.131$ atm. A flask initially contains $N_2O_4(g)$ at 1.000 atm pressure. Calculate the pressures of $NO_2(g)$ and $N_2O_4(g)$ at equilibrium.

**Solution**

Note that the usual steps are followed for this exercise but are condensed:

$$N_2O_4(g) \rightleftarrows 2NO_2(g) \qquad K_p = \frac{P_{NO_2}^{\;2}}{P_{N_2O_4}} = 0.131$$

| *Initial Pressures* | | *Equilibrium Pressures* |
|---|---|---|
| $P_{N_2O_4}{}^0 = 1.000$ atm | Let $x$ atm of $N_2O_4$ react to form $2x$ atm of $NO_2$ to come to equilibrium → | $P_{N_2O_4} = 1.000 - x$ |
| $P_{NO_2}{}^0 = 0$ | | $P_{NO_2} = 2x$ |

$$0.131 = K_p = \frac{P_{NO_2}{}^2}{P_{N_2O_4}} = \frac{(2x)^2}{(1.000 - x)}$$

Multiplying and collecting terms gives the quadratic equation

$$4x^2 + 0.131x - 0.131 = 0$$

where $a = 4$, $b = 0.131$, and $c = -0.131$. Use of the quadratic formula gives

$$x = 0.165 \text{ atm (and an } \textit{extraneous} \text{ root, } x = -0.198 \text{ atm)}$$

The equilibrium pressures are

$$P_{N_2O_4} = 1.000 - x = 0.835 \text{ atm}$$

$$P_{NO_2} = 2x = 2(0.165) = 0.330 \text{ atm}$$

## 1.8 Le Châtelier's Principle

A system will remain at equilibrium unless disturbed in some way. Since certain types of "disturbances" often occur, it is useful to be able to predict in which direction the position of the equilibrium will shift when disturbed. This prediction may be made in terms of Le Châtelier's Principle:

*If a stress is applied to a system at equilibrium, the position of the equilibrium will shift in the direction which reduces the stress.*

Several types of stresses (disturbances) will be considered:

1. Addition or removal of reactants or products.
2. Change in pressure by
   (a) Change in volume.
   (b) Addition of an inert gas.
3. Change in temperature.

The effects of these stresses are explained in the following exercise.

**Sample Exercise 1.11** A 1.000 L vessel contains $N_2(g)$, $H_2(g)$, and $NH_3(g)$ at equilibrium. The reaction is exothermic (heat is produced):

$$3H_2(g) + N_2(g) \rightleftarrows 2NH_3(g)$$

The concentrations are

$$[N_2] = 0.921 \text{ mol/L}$$

$$[H_2] = 0.763 \text{ mol/L}$$

$$[NH_3] = 0.158 \text{ mol/L}$$

**A.** Predict the effect on the equilibrium position of adding 1.000 mol of $H_2$.

**Solution**

The stress is the added $H_2$ which, according to Le Châtelier's Principle, will cause the position of the equilibrium to shift to the right, consuming $H_2$ and relieving the stress. This effect can be rationalized in terms of the reaction quotient. Immediately after addition of $H_2$, the system is no longer at equilibrium. The concentrations at that point can be considered to be initial concentrations:

$$[N_2]_0 = 0.921 \text{ mol/L}$$
$$[NH_3]_0 = 0.158 \text{ mol/L}$$
$$[H_2]_0 = 0.763 \text{ mol/L} + 1.000 \text{ mol/L}$$
$$Q = \frac{[NH_3]_0^2}{[N_2]_0[H_2]_0^3}$$

Since $[H_2]_0$ is greater than the concentration of $H_2$ in the system formerly at equilibrium, $Q$ will be smaller than $K$, and the system will adjust to the right.

Experimental results confirm this conjecture. The new equilibrium concentrations show an increase in $[NH_3]$ and a decrease in $[N_2]$. The details are summarized below.

| *Equilibrium Position I* | | *Equilibrium Position II* |
|---|---|---|
| $[NH_3] = 0.158$ mol/L | add | $[NH_3] = 0.378$ mol/L |
| $[N_2] = 0.921$ mol/L | $\xrightarrow{}$ 1.000 mol $H_2$ | $[N_2] = 0.811$ mol/L |
| $[H_2] = 0.763$ mol/L | | $[H_2] = 1.433$ mol/L |

Note that this shows the shift to the right (toward $NH_3$).

**B.** Predict the effect of adding 1.000 mol of $NH_3(g)$ to the system.

**Solution**

The equilibrium position will shift to the left, since the stress is applied to the right side of the equation ($Q$ will be larger than $K$).

**C.** Predict the effect of suddenly halving the volume (increasing the pressure).

**Solution**

The stress applied is reducing the volume. To relieve this stress, the system will shift in a direction that reduces its own volume. This can be done by a shift to the right, since the right side of the equation involves fewer molecules of gas and thus lower volume.

**D.** Predict the effect of increasing pressure by adding $He(g)$. Assume ideal behavior for all gases.

**Solution**

There will be no effect on the equilibrium position. Ideal gas molecules have negligible volumes. Thus the addition of $He(g)$ does not change the volume, and since He is not involved in the reaction, the concentrations of reacting molecules are unaffected. The equilibrium position remains unchanged.

**E.** Predict the effect of increasing the temperature.

**Solution**

The equilibrium will shift to the left.

This situation is more complicated than those discussed above because the *value of K* changes with temperature. The other disturbances that have been considered (addition or removal of reactants or products and change in pressure) *do not* change the value of $K$. However Le Châtelier's Principle can be used in this case by considering heat to be a reactant or product. The reaction is exothermic, which can be shown by treating heat as a product:

$$3H_2(g) + N_2(g) \rightleftarrows 2NH_3(g) + \text{heat}$$

The temperature is increased by adding heat. Le Châtelier's Principle predicts that the equilibrium position will shift to the left, consuming some of the added heat.

Note that this shift to the left will decrease $[NH_3]$ and increase $[H_2]$ and $[N_2]$ so that $K$ becomes smaller as the temperature is increased. (The opposite effect occurs for an endothermic reaction, where heat may be considered a reactant.)

**TEST 1.10** Consider the reaction

$$2NO_2(g) \rightleftarrows N_2(g) + 2O_2(g)$$

which is exothermic. A vessel contains $NO_2(g)$, $N_2(g)$, and $O_2(g)$ at equilibrium. Predict how each of the following stresses will affect the equilibrium position.

**A.** $NO_2(g)$ is added.
**B.** $N_2(g)$ is removed.
**C.** The volume is halved.
**D.** He($g$) is added.
**E.** The temperature is increased.

## EXERCISES

**1.** Give the equilibrium constant expression for each of the following:
(a) $2H_2(g) + O_2(g) \rightleftarrows 2H_2O(g)$
(b) $SnO_2(s) + 2H_2(g) \rightleftarrows Sn(s) + 2H_2O(g)$
(c) $4NH_3(g) + 7O_2(g) \longrightarrow 4NO_2(g) + 6H_2O(g)$

For exercises 2–5, consider the reaction

$$PCl_3(g) + Cl_2(g) \rightleftarrows PCl_5(g)$$

**2.** Write the equilibrium expression.

**3.** At a certain temperature, the following equilibrium concentrations were observed:

$$[Cl_2] = 2.0 \times 10^{-3} \text{ mol/L}$$

$$[PCl_3] = 3.0 \times 10^{-1} \text{ mol/L}$$

$$[PCl_5] = 6.7 \times 10^{-3} \text{ mol/L}$$

(a) Calculate the value of $K$ at this temperature.
(b) Calculate the value of $K$ for the reaction

$$PCl_5(g) \rightleftarrows PCl_3(g) + Cl_2(g)$$

at the same temperature.

**4.** $Cl_2(g)$, $PCl_3(g)$, and $PCl_5(g)$ are mixed at the following concentrations at the same temperature as in exercise 3:

$$[Cl_2]_0 = 5.0 \times 10^{-4} \text{ mol/L}$$

$$[PCl_3]_0 = 6.3 \times 10^{-2} \text{ mol/L}$$

$$[PCl_5]_0 = 3.8 \times 10^{-3} \text{ mol/L}$$

In which direction will the system shift to reach equilibrium?

**5.** $Cl_2(g)$, $PCl_3(g)$, and $PCl_5(g)$ are in a vessel at equilibrium. In which direction will the position of the equilibrium

$$PCl_3(g) + Cl_2(g) \rightleftarrows PCl_5(g)$$

shift if the total pressure is increased by halving the volume?

**6.** For the reaction

$$2SO_2(g) + O_2(g) \rightleftarrows 2SO_3(g)$$

$K = 4.00 \times 10^2$ L/mol at 750°C. What concentration of $SO_2$ must be in equilibrium with the concentrations $[O_2] = 2.0 \times 10^{-1}$ mol/L and $[SO_3] =$ 3.0 mol/L?

**7.** The synthesis of ammonia,

$$N_2(g) + 3H_2(g) \rightleftarrows 2NH_3(g)$$

is exothermic. The value of $K$ for this equilibrium at 600 K is 4.1 $(L/mol)^2$.
(a) Calculate $K_p$ at 600 K.
(b) Will the value of $K$ be larger or smaller than 4.1 $(L/mol)^2$ at 700 K?

For exercises 8 and 9, consider the reaction

$$H_2(g) + I_2(g) \rightleftarrows 2HI(g)$$

**8.** $H_2(g)$ and $I_2(g)$ were mixed at 490°C and allowed to reach equilibrium. The following equilibrium concentrations were found:

$$[H_2] = 2.00 \text{ mol/L}$$

$$[I_2] = 2.49 \times 10^{-2} \text{ mol/L}$$

$$[HI] = 1.50 \text{ mol/L}$$

(a) Calculate $K$.
(b) Calculate $K_p$.

**9.** In another experiment at 490°C, 1.00 mol of $H_2$, 5.00 mol of $I_2$, and 2.50 mol of HI were mixed in a 1.00 L vessel. When equilibrium was reached, the concentration of $H_2$ was found to be $1.00 \times 10^{-1}$ mol/L. Calculate the equilibrium concentrations of $I_2$ and HI.

**10.** Calculate the value of $K$ at 25°C for the reaction

$$CO_2(g) + H_2(g) \rightleftarrows CO(g) + H_2O(l)$$

if the *equilibrium* concentrations of $CO_2 = H_2 = CO = 0.10$ $M$.

**11.** Consider the reaction

$$CaCO_3(s) \rightleftarrows CaO(s) + CO_2(g)$$

for which $K = 8.0 \times 10^{-3}$ mol/L at 750°C.
(a) 1.00 mol of $CaCO_3(s)$ is sealed into a 5.0 L flask at 750°C. Calculate the equilibrium concentration of $CO_2(g)$.
(b) What fraction (by moles) of the original $CaCO_3(s)$ has decomposed?
(c) How many moles of $CaO(s)$ are produced?
(d) Calculate $K_p$ for this reaction at 750°C.

For exercises 12–14, consider the exothermic reaction

$$2NO(g) + O_2(g) \rightleftarrows 2NO_2(g)$$

A study of this system at 25°C found the equilibrium concentrations to be

$$[NO] = 1.0 \times 10^{-3} \text{ mol/L}$$

$$[O_2] = 1.0 \times 10^{-6} \text{ mol/L}$$

$$[NO_2] = 1.3 \text{ mol/L}$$

**12.** Calculate $K$ at 25°C.

**13.** Calculate $K_p$ at 25°C.

**14.** Predict the effect on the equilibrium position of the following stresses:
(a) Increase the temperature.
(b) Decrease the pressure by increasing the volume.
(c) Increase the pressure by adding Ne($g$).
(d) Add 1.0 mol of $O_2(g)$.

**15.** A 10.0 L flask contains 1.0 mol of $PCl_5(g)$, 0.30 mol of $PCl_3(g)$, and 0.80 mol of $Cl_2(g)$ at equilibrium. Calculate $K$ for the reaction

$$PCl_5(g) \rightleftarrows PCl_3(g) + Cl_2(g)$$

**16.** Consider the reaction

$$2NH_3(g) \rightleftarrows N_2(g) + 3H_2(g)$$

at a temperature where $K = 3.0 \times 10^{-8}$. In an experiment, $1.0 \times 10^{-3}$ mol $NH_3$, $1.0 \times 10^{-1}$ mol $H_2$, and 1.0 mol $N_2$ are mixed in a 5.0 L vessel.
(a) Is this system at equilibrium?
(b) If not, in which direction must the system shift to reach equilibrium?
(c) Define the change needed to reach equilibrium in terms of $x$.
(d) Represent the equilibrium concentrations in terms of the initial concentrations and $x$.

**17.** $N_2O_4(g)$ decomposes to $NO_2(g)$ according to the following equation:

$$N_2O_4(g) \rightleftarrows 2NO_2(g)$$

Pure $N_2O_4(g)$ was placed in a closed flask at 127°C at a pressure of $4.38 \times 10^{-2}$ atm. After the system reached equilibrium, the *total* pressure was found to be $7.43 \times 10^{-2}$ atm. Calculate the value of $K_p$ for this reaction.

### Verbalizing General Concepts

Answer the following in your own words.

**18.** What is chemical equilibrium?

**19.** How does an equilibrium position differ from an equilibrium constant?

**20.** What is the reaction quotient and how is it used?

**21.** What is Le Châtelier's Principle?

**22.** Using Le Châtelier's Principle, state how a change in volume of the reaction vessel of a system at equilibrium affects the equilibrium position.

**23.** The addition of an inert gas to a system at equilibrium has no effect. Explain.

### Multiple Choice Questions

Questions 24–28 deal with the following equilibrium:

$$PCl_5(g) \rightleftarrows PCl_3(g) + Cl_2(g) \qquad \Delta H = 125 \text{ kJ}$$

6.0 mol of $PCl_5(g)$ are placed in an empty 10.0 L reaction vessel at 230°C and the reaction is allowed to come to equilibrium. When equilibrium is reached, an analysis of the mixture indicates that 1.0 mol of $Cl_2(g)$ is present.

**24.** How many moles of $PCl_5(g)$ are present at equilibrium?
(a) 0.0 mol
(b) 1.0 mol
(c) 5.0 mol
(d) 5.6 mol
(e) 6.0 mol

**25.** What is the molar concentration of $Cl_2(g)$ at equilibrium?
(a) 0.1 mol/L
(b) 0.2 mol/L
(c) 0.5 mol/L
(d) 1.0 mol/L
(e) 2.0 mol/L

**26.** The value of the equilibrium constant, $K$, is
(a) $\frac{0.1 \times 0.1}{0.5}$
(b) $\frac{1 \times 1}{5}$
(c) $\frac{1 \times 1}{6}$
(d) $\frac{5}{1 \times 1}$
(e) $\frac{0.5}{0.1 \times 0.1}$

**27.** The numerical value of the equilibrium constant, $K$, for this reaction may be expected to
(a) Increase as temperature is increased to 250°C and pressure is held constant.
(b) Increase as pressure is increased and temperature is held constant.
(c) Increase if some additional $Cl_2$ is injected, temperature and pressure being held constant.
(d) Remain unchanged no matter how conditions are altered.

**28.** Which of the following would surely increase the amount of $PCl_3(g)$ present at equilibrium?
(a) Decrease the volume at constant temperature.
(b) Inject some $Cl_2(g)$ into the system at equilibrium.
(c) Decrease the temperature at constant pressure.
(d) Remove some $PCl_5(g)$ from the system at equilibrium.
(e) Inject some $PCl_5(g)$ into the system at equilibrium.

**29.** Calculate the equilibrium constant, $K_p$ for the following reaction:

$$2NOBr(g) \rightleftarrows 2NO(g) + Br_2(g)$$

given that a 1.0 L vessel was initially filled with 4.0 atm of pure NOBr, and after equilibrium was established, the partial pressure of NOBr gas was 2.5 atm.
(a) 0.45
(b) 0.27
(c) 0.18
(d) 0.75
(e) None of these

**30.** After equilibrium has been attained in the reaction system in question 29, decreasing the total pressure (by increasing the volume) will result in
(a) An increase in the equilibrium constant.
(b) A decrease in the equilibrium constant.
(c) A shift of the equilibrium position to the right.
(d) A shift of the equilibrium position to the left.
(e) None of these.

**31.** In the reaction

$$C(s) + CO_2(g) \xrightleftharpoons{1000\ K} 2CO(g)$$

if one mole of solid carbon is added to the system at equilibrium, the result when equilibrium is reattained will be
(a) To increase the quantity of CO by 2 mol.
(b) To decrease the quantity of CO by 2 mol.
(c) No change.
(d) None of these.

For questions 32–35, consider the reaction

$$Fe_2O_3(s) + 3H_2(g) \rightleftarrows 2Fe(s) + 3H_2O(g) \qquad \Delta H° = 15 \text{ kJ}$$

**32.** The equilibrium expression is

(a) $K = \dfrac{[Fe]^2[H_2O]^3}{[Fe_2O_3][H_2]^3}$

(b) $K = \dfrac{1}{[H_2]^3}$

(c) $K = \dfrac{[H_2O]^3}{[H_2]^3}$

(d) None of these

**33.** The equilibrium concentrations under certain conditions were found to be

$$[H_2O] = 1.0 \text{ mol/L}$$

$$[H_2] = 2.5 \text{ mol/L}$$

Calculate $K$.

(a) 0.40

(b) 0.064

(c) 15.6

(d) Insufficient information to calculate

**34.** $K_p$ is

(a) Less than $K$

(b) Greater than $K$

(c) Equal to $K$

(d) Insufficient information to predict

**35.** Use the following choices:

(a) Shift left

(b) Shift right

(c) No change

(d) Not possible to predict

to indicate the effect of each of the following stresses on the position of this system at equilibrium:

(1) Decrease the volume of the container.

(2) Add $Fe_2O_3(s)$.

(3) Remove $H_2O(g)$.

(4) Increase the temperature.

**36.** For the reaction

$$4NH_3(g) + 5O_2(g) \rightleftarrows 4NO(g) + 6H_2O(g)$$

$\Delta H°$ is negative. The position of this equilibrium would be shifted to the left by

(a) Removing $NO(g)$.

(b) Adding $O_2(g)$.

(c) Increasing the pressure by decreasing the volume of the container.

(d) Decreasing the temperature.

(e) None of these.

**37.** Calculate the equilibrium constant for the following reaction:

$$2A(g) \rightleftarrows B(g) + 3C(g)$$

given that a 1.0 L vessel was initially filled with 5.0 atm of pure A and the partial pressure of gas A was found to be 3.5 atm at equilibrium.

(a) 1.1 $atm^2$

(b) 0.70 $atm^2$

(c) 3.8 $atm^2$

(d) 0.48 $atm^2$

(e) None of these

**38.** Increasing the volume of the reaction vessel in question 37 would have what effect?

(a) Increase the equilibrium constant.

(b) Force the reaction to proceed to the right.

(c) Force the product ratio [C]/[B] to be greater than 3.
(d) Only (a) and (b) are correct.
(e) (a), (b), and (c) are correct.

**39.** Consider the following equilibrium system:

$$SnCl_2(s) + Cl_2(g) \rightleftarrows SnCl_4(l) + 195 \text{ kJ}$$

The position of this equilibrium could be driven to the *right* by
(a) Compressing the system in the reaction chamber.
(b) Increasing the temperature in the reaction chamber.
(c) Adding more $SnCl_4(l)$ to the reaction mixture.
(d) Adding more $SnCl_2(s)$ to the reaction mixture.
(e) All of the above would drive the reaction farther to the right.

**40.** Given the following equation:

$$A + B \rightleftarrows 3C + D$$

where initial concentrations of A and B are each equal to 1.2 mol/L with no C or D present. At equilibrium, the concentration D is found to be 0.30 mol/L. Calculate $K$.
(a) 0.27 (b) 0.33
(c) 0.010 (d) 0.030
(e) None of these

**41.** For the gas phase oxidation of CO to $CO_2$ by means of $O_2$:

$$2CO + O_2 \rightleftarrows 2CO_2$$

the equilibrium constant expression is
(a) $K = \dfrac{[CO_2]^2}{[CO][O_2]}$ (b) $K = \dfrac{[CO_2]^2}{[CO]^2[O_2]}$
(c) $K = \dfrac{[CO]^2[O_2]}{[CO_2]}$ (d) $K = \dfrac{[CO]^2[O_2]}{[CO_2]^2}$
(e) $K = \dfrac{[CO][O_2]}{[CO_2]}$

**42.** Two moles of $NH_3$ gas are introduced into a previously evacuated 1.0 L container in which it partially dissociates at high temperature:

$$2NH_3(g) \rightleftarrows N_2(g) + 3H_2(g)$$

At equilibrium, 1.0 mol of $NH_3(g)$ remains. What is the equilibrium constant for the above reaction?
(a) 0.42 (b) 0.75
(c) 1.5 (d) 1.7
(e) None of these

**43.** Which of the following affects the *value* of $K$ for gaseous reactions?
(a) Change in pressure of reactants or products.
(b) Change in concentration of reactants or products.
(c) Change in temperature.
(d) Change of volume of the container.
(e) None of these.

For questions 44–48, consider the equilibrium

$$H_2O(g) + CO(g) \rightleftarrows H_2(g) + CO_2(g) \qquad \Delta H^\circ = -40 \text{ kJ}$$

Use the following choices
(a) Shift to the right (b) Shift to the left
(c) No change (d) Not enough information to predict

to give the effect of each of the following stresses on the position of the equilibrium for this reaction:

**44.** 1.0 mol of $H_2(g)$ is added.

**45.** 1.0 mol of $CO(g)$ is removed.

**46.** 1.0 mol of $He(g)$ is added.

**47.** The pressure is doubled by halving the volume.

**48.** The temperature is decreased.

**49.** In which reaction will a decrease in volume at constant temperature favor formation of the products?
(a) $CaCO_3(s) \rightleftarrows CaO(s) + CO_2(g)$ (b) $H_2(g) + Cl_2(g) \rightleftarrows 2HCl(g)$
(c) $2NO(g) + O_2(g) \rightleftarrows 2NO_2(g)$ (d) $COCl_2(g) \rightleftarrows CO(g) + Cl_2(g)$

**50.** Given the equilibrium equation

$$A_2(g) + B_2(g) \rightleftarrows 2AB(g)$$

where the equilibrium concentrations in mol/L are

$$[A_2] = 0.40$$

$$[B_2] = 0.40$$

$$[AB] = 1.20$$

what is the value of the equilibrium constant?
(a) 0.11 (b) 0.13
(c) 1.8 (d) 9.0

**51.** For the equilibrium system $N_2(g) + 2O_2(g) \rightleftarrows 2NO_2(g)$, where the forward reaction is endothermic, which of the following statements is *not* true?
(a) A decrease in volume will shift the position of the equilibrium to the right.
(b) An increase in temperature will shift the equilibrium to the left.
(c) The equilibrium expression is $K = [NO_2]^2/[N_2][O_2]^2$
(d) A decrease in the concentration of $N_2$ will shift the position of the equilibrium to the left.

**52.** For the reaction $2CO(g) + O_2(g) \rightleftarrows 2CO_2(g)$, the initial concentrations for CO, $O_2$, and $CO_2$ are each 0.10 mol/L. $K = 4 \times 10^{-16}$ at 500°C. To establish equilibrium, the system will
(a) Proceed to the right. (b) Proceed to the left.
(c) Not change. (d) Cannot tell; insufficient information.

**53.** The equilibrium constants $K_I$ and $K_{II}$ for the reactions

$$\text{I. } N_2(g) + 3H_2(g) \rightleftarrows 2NH_3(g)$$

$$\text{II. } \tfrac{1}{2}N_2(g) + \tfrac{3}{2}H_2(g) \rightleftarrows NH_3(g)$$

are related as
(a) $K_I = K_{II}$ (b) $K_I = \dfrac{1}{K_{II}}$
(c) $K_{II} = \sqrt{K_I}$ (d) $K_I = \sqrt{K_{II}}$
(e) None of these

**Note:** The following exercises require the solution of quadratic equations.

**54.** Consider the reaction

$$PCl_3(g) + Cl_2(g) \rightleftarrows PCl_5(g)$$

at a temperature where $K = 11$ L/mol. If $1.0 \times 10^{-3}$ mol of $Cl_2(g)$, 1.26 $\times 10^{-1}$ mol $PCl_3(g)$, and $7.6 \times 10^{-3}$ mol of $PCl_5(g)$ are mixed in a 2.0 L vessel, calculate the equilibrium concentrations of $Cl_2(g)$, $PCl_3(g)$, $PCl_5(g)$.

**55.** 1.00 mol of $N_2O_4(g)$ is placed in a 10.0 L vessel and allowed to reach equilibrium. The reaction of interest is

$$N_2O_4(g) \rightleftarrows 2NO_2(g)$$

where $K = 8.1 \times 10^{-2}$ mol/L. Calculate the equilibrium concentrations of $N_2O_4(g)$ and $NO_2(g)$.

For questions 56 and 57, consider the reaction

$$N_2(g) + 3H_2(g) \rightleftarrows 2NH_3(g)$$

where $K_p = 1.7 \times 10^{-3}$ atm$^{-2}$ at 600 K.

**56.** 0.10 atm of $N_2$ is mixed with 0.50 atm of $H_2$ in a 1.0 L flask at 600 K. Calculate the equilibrium pressures of $N_2(g)$, $H_2(g)$, and $NH_3(g)$.

**57.** 0.0200 atm of $H_2(g)$, 0.0800 atm of $N_2(g)$, and 0.100 atm of $NH_3(g)$ are placed in a 1.00 L flask at 600 K.
(a) In which direction will the system shift to come to equilibrium?
(b) Calculate the equilibrium pressures of $N_2(g)$, $H_2(g)$, and $NH_3(g)$.

For questions 58 and 59, consider the reaction

$$H_2(g) + I_2(g) \rightleftarrows 2HI(g)$$

at a temperature where $K = 38.6$.

**58.** 1.800 mol of $H_2$, 1.800 mol of $I_2$, and 2.600 mol of HI are mixed in a 2.000 L vessel. Calculate the equilibrium concentrations of $H_2$, $I_2$, and HI.

**59.** 1.00 mol of $I_2$ is added to the system at equilibrium in question 58.
(a) In which direction does the position of the equilibrium shift?
(b) Calculate the new equilibrium concentrations.

**60.** Consider the equilibrium mixture from question 15. Now 1.00 mol of $PCl_5(g)$ is added to this mixture in the 10.0 L vessel. Calculate the new equilibrium concentrations of $Cl_2$, $PCl_3$, and $PCl_5$.

**61.** Consider the reaction

$$2HI(g) \rightleftarrows H_2(g) + I_2(g)$$

at 620°C where $K = 1.63 \times 10^{-3}$. In an experiment, 0.600 mol of $H_2(g)$, 0.300 mol of $I_2(g)$, and 0.800 mol of $HI(g)$ are added to a 2.0 L vessel. Calculate the equilibrium concentrations of HI, $H_2$, and $I_2$.

**62.** Consider the reaction

$$2ClO(g) \rightleftarrows Cl_2(g) + O_2(g)$$

at a temperature where $K = 6.4 \times 10^{-3}$. In an experiment, $1.00 \times 10^{-1}$ mol of $ClO(g)$, 1.00 mol of $O_2(g)$, and $1.00 \times 10^{-2}$ mol of $Cl_2(g)$ were mixed in a 4.00 L flask.
(a) Is the system at equilibrium?
(b) If not, in which direction must the system proceed to reach equilibrium?
(c) Calculate the concentrations of $ClO(g)$, $Cl_2(g)$, and $O_2(g)$ at equilibrium.

# 2 Acid-Base Equilibria

## CHAPTER OBJECTIVES

1. Define acid, base, $K_a$, $K_b$, pH, and buffer.
2. Understand the concept of acid strength.
3. Recognize the major species in solutions containing acids and/or bases.
4. Calculate the pH of a solution containing an acid and/or a base.
5. Learn to deal with buffered solutions including:
   (a) Calculation of the pH of buffered solutions.
   (b) Calculation of the change in pH when $OH^-$ or $H^+$ is added to a buffered solution.
6. Learn to approach complicated problems in a systematic way.
7. Learn to make reasonable assumptions to simplify the calculations involved in acid-base problems.

## 2.1 Introduction

Acids and bases in aqueous solution are most often defined in terms of the **Brønsted-Lowry model:** An acid is a proton ($H^+$) donor and a base is a proton acceptor. In general terms, the reaction that occurs when an acid is dissolved in water can be represented as

$$\underset{\text{Acid}}{HA(aq)} + \underset{\text{Base}}{H_2O(l)} \rightleftarrows \underset{\text{Conjugate acid}}{H_3O^+(aq)} + \underset{\text{Conjugate base}}{A^-(aq)} \tag{2.1}$$

where "*aq*" means that the substance is hydrated (water molecules are attached).

Note that this is an equilibrium system that involves competition for the proton between two bases, $H_2O$ and $A^-$. If $H_2O$ is a much stronger base than $A^-$, it will combine with most of the $H^+$ and the position of the equilibrium will be to the right. If $A^-$ is a much stronger base than $H_2O$, the equilibrium position will be to the left.

The equilibrium constant expression is

$$K_a = \frac{[H_3O^+][A^-]}{[HA]} = \frac{[H^+][A^-]}{[HA]}$$

where $K_a$ is called the **dissociation constant** for the acid. Note the following important points:

1. The symbols "$H^+(aq)$" or simply "$H^+$" are often used to represent the hydrated proton, $H_3O^+(aq)$.

2. The condensed phase, $H_2O$, the concentration of which remains essentially constant, is not included in the $K_a$ equilibrium expression. Because $[H_2O(l)]$ is not included, the form of the $K_a$ equilibrium expression is the same as if the reaction were a simple dissociation:

$$HA(aq) \rightleftarrows H^+(aq) + A^-(aq)$$

The water molecules are very important in causing an acid to "dissociate" (the $H^+$ ions are transferred to $H_2O$ molecules). However, since $[H_2O(l)]$ is constant and does not enter into the equilibrium calculations, the reaction of an acid in water is often represented as a simple dissociation, neglecting the role of water.

3. The reaction corresponding to $K_a$ *always* involves formation of $H^+$ and a conjugate base. The conjugate base is everything left after the removal of $H^+$ from the acid. Knowing this fact allows one to write the $K_a$ reaction even for an unfamiliar acid.

## Acid Strength

The strength of an acid is determined by the position of the "dissociation" equilibrium. A **strong acid** dissociates completely in water. A **weak acid**, on the other hand, dissociates only to a slight extent. Several general statements can be made concerning strong and weak acids:

*Strong Acid*

1. $K_a$ is very large.
2. $[H^+] = [HA]_0$, where $[HA]_0$ represents the number of moles of acid dissolved per liter of solution.
3. $A^-$ is a weak base relative to $H_2O$.

The common strong acids are hydrochloric, $HCl(aq)$; nitric, $HNO_3(aq)$; and sulfuric, $H_2SO_4(aq)$.

*Weak Acid*

1. $K_a$ is very small.
2. $[H^+] \ll [HA]_0$. ($\ll$ means "much less than")
3. $A^-$ is a stronger base than $H_2O$ and remains for the most part bound to the proton.

It is very important to focus on the relationship between the strength of an acid and the strength of its conjugate base. A strong acid implies a very weak conjugate base (one with very small affinity for $H^+$ compared to $H_2O$). A weak acid, on the other hand, has a conjugate base that is strong compared to $H_2O$.

---

**TEST 2.1**

**A.** Describe in words what reaction is associated with $K_a$ and what $K_a$ means.

**B.** Write the $K_a$ expression for

1. HCN
2. $C_6H_5NH_3^+$
3. $Al(OH_2)_6^{3+}$

Hint: Although acids 2 and 3 may be unfamiliar to you, simply apply the definition of a Brønsted acid.

**C.** The $K_a$ for HCN is $6.2 \times 10^{-10}$. State the value of $K$ for the reaction

$$H^+(aq) + CN^-(aq) \rightleftarrows HCN(aq)$$

**D.** Consider the following acids:

| *Acid* | $K_a$ |
|---|---|
| $HIO_3$ | $1.7 \times 10^{-1}$ |
| $HNO_2$ | $4.0 \times 10^{-4}$ |
| HF | $7.2 \times 10^{-4}$ |
| HOCl | $3.5 \times 10^{-8}$ |

1. Which of the anions $IO_3^-$, $NO_2^-$, $F^-$, and $OCl^-$ is the strongest base? (Hint: Recall the relationship between acid strength and the strength of the conjugate base.)
2. Order these anions according to decreasing base strength.
3. Order the acids according to increasing acid strength.

## 2.2 General Principles

### Water as an Acid and Base

A substance is said to be **amphoteric** if it can react as either an acid or a base. Water is the most common amphoteric substance. This property can most clearly be seen in the **autoionization** of water:

$$H_2O(l) + H_2O(l) \rightleftarrows H_3O^+(aq) + OH^-(aq)$$

In this reaction, one molecule of water furnishes a proton, behaving as an acid, and a second molecule accepts a proton, behaving as a base. The equilibrium expression for this process is

$$[H_3O^+][OH^-] = [H^+][OH^-] = K_w$$

At 25°C, experiment shows that

$$[H^+] = [OH^-] = 1.0 \times 10^{-7}\ M$$

Substituting these concentrations into the equilibrium constant expression gives a value of $1.0 \times 10^{-14}$ for $K_w$. Thus in any aqueous solution at 25°C, the *product* of the $[H^+]$ and $[OH^-]$ must be $1.0 \times 10^{-14}$ at equilibrium.

Note that there are an infinite number of combinations of $[H^+]$ and $[OH^-]$ that when multiplied together will give $1.0 \times 10^{-14}$. For example, $(1 \times 10^{-5})(1 \times 10^{-9}) = (1 \times 10^{-4})(1 \times 10^{-10}) = (5 \times 10^{-3})(2 \times 10^{-12}) = 1 \times 10^{-14}$. Recall from chapter 1 that there are an infinite number of equilibrium positions but only one equilibrium constant at a given temperature.

**Sample Exercise 2.1** In a certain acidic solution at 25°C, the $[H^+] = 1.0 \times 10^{-2}\ M$. What is the $[OH^-]$ in this solution?

**Solution**

We know that

$$[H^+][OH^-] = K_w = 1.0 \times 10^{-14}$$

Since $[H^+] = 1.0 \times 10^{-2}$,

$$[H^+][OH^-] = (1.0 \times 10^{-2})[OH^-] = 1.0 \times 10^{-14}$$

$$[OH^-] = \frac{1.0 \times 10^{-14}}{1.0 \times 10^{-2}} = 1.0 \times 10^{-12}\ M$$

## The pH Scale

Because the concentration of $H^+$ in most aqueous solutions is small (for example, it is $1.0 \times 10^{-7}$ $M$ in pure water), a quantity, called **pH**, is defined to designate the $[H^+]$, where

$$pH = -\log[H^+]$$

A similar type of notation is used for other quantities, for example,

$$pOH = -\log[OH^-]$$

$$pK = -\log K$$

**Sample Exercise 2.2**

**A.** Calculate the pH of pure water.

**Solution**

The $[H^+] = 1.0 \times 10^{-7}$ $M$, so

$$pH = -\log[H^+] = -\log(1.0 \times 10^{-7}) = -(-7.00) = 7.00$$

**B.** Calculate the pOH of pure water.

**Solution**

The $[OH^-] = 1.0 \times 10^{-7}$ $M$

$$pOH = -\log[OH^-] = -\log(1.0 \times 10^{-7}) = 7.00$$

**TEST 2.2**

**A.** Calculate the pH of a solution in which the $[OH^-] = 2.0 \times 10^{-4}$ $M$.

**B.** Calculate the pOH of a solution containing $5.0 \times 10^{-3}$ $M$ $H^+$.

## Solving Acid-Base Problems: Some General Considerations

Acid-base problems tend to be more complicated than gas-phase equilibrium problems because aqueous solutions typically contain several components. It is therefore important to approach acid-base problems systematically and with care.

One of the most important steps in solving an acid-base problem is deciding which components are important and which can be ignored. In fact, knowing what simplifications can be made lies at the heart of the methods for solving many types of chemistry problems. Chemical systems tend to be very complex, and many are impossible to treat exactly. Thus, in order to obtain some kind of answer, reasonable approximations must be made. Much of chemistry involves the art of making intelligent approximations.

One way to approach chemistry is to memorize every problem you can find and then hope that test questions are exactly the same. This is a poor strategy that is likely to produce only frustration and anger. Instead of wasting time memorizing problems, homework should be used to practice the application of chemical concepts. In approaching a chemistry problem, do not try to force it into the mold of some previously memorized solution. Do not worry if you cannot immediately see the entire solution to the problem. Approach the problem in a step-by-step fashion, and *let the problem guide you* toward a solution. Don't fight the problem; it will often win.

In treating acid-base problems, the emphasis will be on the need to be systematic, yet flexible, and to look at each problem with an open mind. The methods illustrated for solving these problems may seem awkward at first, but they will pay dividends as the problems become more complicated. Remember, the key is to read the problem carefully and let the information given be the guide to the solution.

## 2.3 Calculating the pH for Solutions Containing Strong Acids

Recall that a strong acid is an acid that is completely dissociated in water. Methods for dealing with strong acids will be developed by doing the following sample exercise.

**Sample Exercise 2.3**

Calculate the $[H^+]$ and the pH in a 1.0 *M* solution of HCl.

**Solution**

The following steps should always be used when doing an acid-base problem:

1. Note the substances used to prepare the solution (HCl and $H_2O$ in this case) and then list the *major species* in the solution. This is a critical step; the ground rules must be carefully established. The major species:
   (a) Are the components of the solution present in relatively large amounts.
   (b) Are written as they actually occur in solution.

   To illustrate: What components are present in 1.0 *M* HCl?

Since HCl is a strong acid (essentially completely dissociated), the solution contains no HCl molecules; rather, it contains $H^+$ and $Cl^-$ ions. The label "1.0 *M* HCl" tells how the solution was prepared (1.0 mol HCl(*g*) per liter of solution), not that it actually contains 1.0 mol/L of HCl molecules. The solution also contains $H_2O(l)$, which dissociates to a slight extent to produce $H^+$ and $OH^-$.

Thus the solution contains

$$H^+, Cl^-, OH^-, H_2O$$

Which of these are major species?

Note that $H_2O$, $H^+$, and $Cl^-$ are present in relatively large quantities and therefore are classified as major species. Since $OH^-$ is present in very small quantities, it will be ignored for now.

2. Write reactions for the substances that produce (or consume) $H^+$. In this case, there are two sources of $H^+$:
   (a) $HCl(aq) \longrightarrow H^+(aq) + Cl^-(aq)$
   This has already been assumed to be complete in considering major species.
   (b) $H_2O(l) \rightleftarrows H^+(aq) + OH^-(aq)$

3. Is there a dominant source of $H^+$ in the solution?

The dissociated HCl produces 1.0 *M* $H^+$. The only other acid is $H_2O$. How important is $H_2O$ as a source of $H^+$? In pure water, $H_2O$ produces $1.0 \times 10^{-7}$ *M* $H^+$. In this problem, because of the large amount of $H^+$ from the HCl, the water equilibrium

$$H_2O(l) \rightleftarrows H^+(aq) + OH^-(aq)$$

will be shifted to the left so that the contribution to $[H^+]$ from the autoionization of $H_2O$ will be much smaller than $1.0 \times 10^{-7}$ *M*. Thus $[H^+]_{total} = 1.0$ *M* (from HCl) plus a very small contribution from $H_2O$, which means that, for all practical purposes,

$$[H^+] = 1.0\ M$$

Note that this problem could be very complicated but is greatly simplified because $H_2O$ can be ignored as a source of $H^+$. *Remember*: Intelligent approximations are the key to solving acid-base problems.

Since $[H^+] = 1.0$ *M*,

$$pH = -\log[H^+] = -\log(1.0) = 0$$

The pH of 1.0 *M* HCl is 0. Note that a pH of zero does *not* mean that the $[H^+] = 0$.

**TEST 2.3**

**A.** Calculate the pH of 0.10 *M* $HNO_3$.

1. List the materials used to prepare this solution and the major species in solution.
2. Indicate the major and minor sources of $H^+$. (Write the reactions that produce $H^+$.)
3. Indicate the approximations to be made.
4. Calculate the pH.

**B.** Calculate the pH of $1.0 \times 10^{-10}$ *M* HCl.

1. List the species in solution.
2. Indicate the sources of $H^+$. (Write reactions.)
3. Choose the dominant source of $H^+$.
4. Calculate the pH.

## 2.4 Calculating the pH for Solutions Containing Weak Acids

Now consider the weak acid problem. This situation is more complicated than that for a strong acid. Since a weak acid is not completely dissociated, an equilibrium calculation must be performed to find the $[H^+]$. A weak acid can be recognized by its small $K_a$ value.

**Sample Exercise 2.4** Calculate (a) the $[H^+]$ and the $[F^-]$, (b) the pH, and (c) the percentage dissociation in a 1.00 *M* solution of hydrofluoric acid, $HF(aq)$. The $K_a$ for $HF(aq)$ is $7.2 \times 10^{-4}$.

**Solution**

This problem will be solved by using a series of steps that will be summarized at the end.

1. First list the major species: HF(*aq*) and $H_2O$.

   Note that the acid present is represented as HF rather than as $H^+$ and $F^-$. HF is a weak acid (small $K_a$) and will be present primarily as HF.

2. Determine which of the major species contribute to the $[H^+]$ and write the equations for the reactions that produce $H^+$.

   Both $H_2O$ and HF are acids:

$$H_2O(l) \rightleftarrows H^+(aq) + OH^-(aq) \qquad K_w = 1.0 \times 10^{-14}$$

$$HF(aq) \rightleftarrows H^+(aq) + F^-(aq) \qquad K_a = 7.2 \times 10^{-4}$$

   From the magnitudes of the equilibrium constants, it can be seen that although HF is a weak acid, it is much stronger than $H_2O$. Thus the dissociation of HF will make the dominant contribution to the $[H^+]$.

   *Use the size of the equilibrium constants to decide which acid is dominant.*

3. Write the equilibrium expression for the dominant acid:

$$HF(aq) \rightleftarrows H^+(aq) + F^+(aq) \qquad K_a = \frac{[H^+][F^-]}{[HF]}$$

4. List the *initial concentrations* (the concentrations before any acid dissociation occurs) of all species in the dominant equilibrium:

$$[HF]_0 = 1.00\ M$$

$$[F^-]_0 = 0$$

$$[H^+]_0 = 10^{-7} \text{ from water, which we will assume can be ignored}$$

   Thus

$$[H^+]_0 \approx 0 \ (\approx \text{ indicates an approximation})$$

5. Define $x$, the change needed to achieve equilibrium.

   Let $x$ represent the HF that dissociates to bring the system to equilibrium (use units of mol/L). Refer to the balanced equation in step 3, and note that $x$ mol/L of HF will produce $x$ mol/L of $H^+$ and $x$ mol/L of $F^-$.

6. Write the equilibrium concentrations in terms of the initial concentrations and $x$:

$$[HF] = [HF]_0 - x = 1.00 - x$$

$$[F^-] = [F^-]_0 + x = 0 + x = x$$

$$[H^+] = [H^+]_0 + x = 0 + x = x$$

7. Substitute the equilibrium concentrations into the equilibrium expression:

$$K_a = 7.2 \times 10^{-4} = \frac{[H^+][F^-]}{[HF]} = \frac{(x)(x)}{1.00 - x} = \frac{x^2}{1.00 - x}$$

8. Simplify the expression by neglecting $x$, if possible. Then solve for $x$.

   For example, note that expansion of the equation in step 7 leads to a quadratic equation. To solve this equation exactly requires use of the quadratic formula. This process can be avoided by making an intelligent approximation. Notice that $K_a = 7.2 \times 10^{-4}$, which indicates that HF does not dissociate to a large extent. Thus $x$ should be a small number. The term "$1.00 - x$" appears in the denominator. If $x$ is very small compared to 1.00, then

$$1.00 - x \approx 1.00$$

This will greatly simplify the math:

$$7.2 \times 10^{-4} = \frac{x^2}{1.00 - x} = \frac{x^2}{1.00}$$

$$x \approx (1.00)(7.2 \times 10^{-4})$$

$$x \approx \sqrt{7.2 \times 10^{-4}} = 2.7 \times 10^{-2}$$

9. Check the validity of the approximation made in step 8.

   In step 8, the assumption was made that $x$ is small compared to 1.00. Is this justified? From the above calculation,

$$x = 2.7 \times 10^{-2}$$

   Subtracting $x$ from 1.00 gives

$$1.00 - x = 1.00 - 0.027 = 0.97$$

   Thus, $1.00 - x$ is not very different from 1.00, so that

$$[HF] \approx [HF]_0$$

   Usually it can be assumed that the equilibrium concentration, [HA], of a weak acid (which dissociates only to a slight extent) is the same as the initial concentration, $[HA]_0$. That is,

$$[HA] = [HA]_0 - x \approx [HA]_0$$

   The best way to decide if this approximation is valid in a given case is to do the problem the easy way (*assume* the approximation *is* valid) and then check to see how large the calculated value of $x$ is, relative to $[HA]_0$. We will use the convention that if $x/[HA]_0 \times 100$ *is less than 5%*, the approximation $[HA]_0 - x \approx [HA]_0$ will be considered valid. (A 5% error in the value of [HA] produces approximately a 2–3% error in the calculated $[H^+]$.)

   In the present problem,

$$\frac{x}{1.00} \times 100 = \frac{2.7 \times 10^{-2}}{1.00} \times 100 = 2.7\%$$

   The error is less than 5%, so the approximation is considered to be valid.

10. Calculate the $[H^+]$ and $[F^-]$.

    Note that $x = 2.7 \times 10^{-2}$ mol/L. In step 4, the assumption was made that water makes a negligible contribution to the final $[H^+]$. Is this assumption valid? Yes. The $[H^+]$ from HF ($2.7 \times 10^{-2}$) is more than $10^5$ times the *maximum* that could be contributed to water ($10^{-7}$ mol/L). Thus from the expression for the equilibrium concentration in step 6,

$$[H^+] = x = 2.7 \times 10^{-2} \text{ mol/L}$$

    Also

$$[F^-] = x = 2.7 \times 10^{-2} \text{ mol/L}$$

11. Calculate the pH.

    From step 10,

$$[H^+] = 2.7 \times 10^{-2} \text{ mol/L}$$

$$pH = -\log[H^+] = -\log(2.7 \times 10^{-2}) = 1.57$$

12. Calculate the percentage dissociation of the HF in this solution.

The *percentage dissociation* of a weak acid is defined as follows:

$$\% \text{ dissociation} = \frac{\text{no. of mol/L of HA dissociated}}{\text{no. of original mol/L of HA}} \times 100$$

In this problem,

$$x = \text{no. of mol/L of HF dissociated} = 2.7 \times 10^{-2}$$

Thus

$$\% \text{ dissociation} = \frac{2.7 \times 10^{-2} \text{ mol/L}}{1.00 \text{ mol/L}} \times 100 = 2.7\%$$

## Summary of the Steps

The important steps in attacking a weak acid problem are

1. List the major species in solution, as they exist in solution. Strong acids are written as completely dissociated. Weak acids are written as HA.
2. Write the balanced reactions for major species that produce $H^+$, and determine the dominant source of $H^+$. This can be done in a typical case by looking at the values of the equilibrium constants. In almost every case, one acid will clearly be dominant.
3. Write the equilibrium expression for the dominant acid.
4. List the initial concentrations of all species involved in the dominant equilibrium.
5. Define $x$, the change required to reach equilibrium.
6. Write the equilibrium concentrations in terms of the initial concentrations and $x$.
7. Substitute the equilibrium concentrations in the expression for $K_a$.
8. Simplify the expression by neglecting $x$ where possible. That is, assume that $[HA] = [HA]_0 - x \approx [HA]_0$. Then solve for $x$.
9. Check the validity of the assumption made in step 8. (Use the 5% rule.)
10. Calculate the $[H^+]$ from the definition of $[H^+]$ in step 6.
11. Calculate the pH.

**TEST 2.4** A. Consider a 0.100 *M* solution of hypochlorous acid (HOCl), which has a $K_a$ of $3.5 \times 10^{-8}$.

1. List the *major* species in this solution.
2. Write balanced equations for dissociation of the acids. Decide which of these will be the dominant contribution to the $[H^+]$.
3. Write the equilibrium constant expression for the dominant equilibrium.
4. List the initial concentrations for the species in this equilibrium.
5. Define $x$.
6. Write the equilibrium concentrations in terms of $x$ and the initial concentrations.
7. Substitute the equilibrium concentrations into the $K_a$ expression.
8. Solve for $x$ using the "easy method."
9. Decide whether the assumption made in step 8 is valid.
10. Calculate the $[H^+]$ and the pH.
11. Calculate the $[OCl^-]$ at equilibrium.

12. Calculate the percent dissociation of HOCl in this solution.
13. Calculate the $[OH^-]$ in this solution.

**B.** Calculate the pH of a 0.50 *M* solution of boric acid ($K_a = 5.8 \times 10^{-10}$). Use the formula $H_3BO_3$ to represent the formula of boric acid. Remember to be systematic. Follow the steps given above.

---

**Sample Exercise 2.5**

0.050 mol of a weak acid (HA) is dissolved in enough water to make 1.0 L of solution. The pH of this solution is found to be 3.50. What is the value of $K_a$ for this weak acid?

**Solution**

This problem is different from those considered previously. In this case, the pH is given (which will allow the calculation of the $[H^+]$) and the value of $K_a$ is to be calculated.

One of the secrets to solving acid-base problems is always to use the same system to approach every problem. Thus, even though this problem appears to be quite different from previous problems, the procedure will use the same steps.

1. The major species in solution are

$$HA, H_2O$$

2. These are both acids:

$$H_2O \rightleftarrows H^+ + OH^-$$
$$HA \rightleftarrows H^+ + A^-$$

Since the given pH is significantly less than 7, the weak acid, HA, must be more important than $H_2O$ as a contributor to $[H^+]$.

3. The $K_a$ expression for HA is

$$K_a = \frac{[H^+][A^-]}{[HA]}$$

4. The initial concentrations are

$$[HA]_0 = 0.050\ M$$
$$[A^-]_0 = 0$$
$$[H^+]_0 \approx 0 \quad \text{(ignoring the contribution of } H_2O\text{)}$$

5. Determine the change needed to reach equilibrium. The change required is usually called $x$. However, in this case the pH is given. Calculate the $[H^+]$ in this solution by taking the antilog (INV and the LOG on most calculators) of $-$pH:

$$[H^+] = \text{antilog}(-\text{pH}) = 10^{-\text{pH}}$$

In this case, the pH = 3.50. Take the antilog of $-3.50$:

$$[H^+] = 10^{-3.50} = 3.16 \times 10^{-4}\ M$$

This $[H^+]$ has resulted from dissociation of HA. Thus $3.16 \times 10^{-4}$ mol/L of HA must have dissociated for the system to reach equilibrium. The change to reach equilibrium is thus $3.16 \times 10^{-4}$ mol/L.

6. The equilibrium concentrations are

$$[HA] = [HA]_0 \text{ plus change} = 0.050 - 3.16 \times 10^{-4} \approx 0.050$$
$$[A^-] = [A^-]_0 \text{ plus change} = 0 + 3.16 \times 10^{-4}$$
$$[H^+] = [H^+]_0 \text{ plus change} = 0 + 3.16 \times 10^{-4}$$

Our unknown in this case is the equilibrium constant. (Reread the question if necessary.)

7. Substitute the equilibrium concentrations into the $K_a$ expression:

$$K_a = \frac{[H^+][A^-]}{[HA]} = \frac{(3.16 \times 10^{-4})(3.16 \times 10^{-4})}{0.050 - 3.16 \times 10^{-4}} \approx \frac{(3.16 \times 10^{-4})^2}{5.0 \times 10^{-2}} = 2.0 \times 10^{-6}$$

This is the value of $K_a$ required by the problem.

## 2.5 Calculations for Solutions of Weak Acids Containing a Common Ion

This type of problem deals with a solution containing a weak acid and a soluble salt containing the anion (conjugate base) of the weak acid. The acid and salt thus have the same anion.

Sodium and potassium salts are usually quite soluble in water and are often used. It is important to recognize that when a sodium or potassium salt (ionic compound) dissolves in water, it can be assumed to be completely dissociated. Thus a 1.0 *M* NaCl solution really contains 1.0 *M* $Na^+$ and 1.0 *M* $Cl^-$; 5.0 *M* KF contains 5.0 *M* $K^+$ and 5.0 *M* $F^-$, etc. This is something you must remember. It will be important in solving many types of chemistry problems.

**Sample Exercise 2.6** Calculate the $[H^+]$ in a solution containing 1.00 *M* hydrofluoric acid (HF, $K_a = 7.2 \times 10^{-4}$) and 0.50 *M* sodium fluoride (NaF).

**Solution**

1. The major species in solution are

   HF

   dissolved NaF ($Na^+$, $F^-$)

   $H_2O$

2. Consider the acid-base properties of each component:
   (a) HF is a weak acid ($HF \rightleftarrows H^+ + F^-$).
   (b) $F^-$ is the conjugate base of HF.
   (c) $Na^+$ has no acid or base properties.
   (d) $H_2O$ is a much weaker acid than HF ($K_a$ for HF $\gg K_w$).

3. Thus the HF dissociation equilibrium will determine the $[H^+]$:

$$HF(aq) \rightleftarrows H^+(aq) + F^-(aq) \qquad K_a = \frac{[H^+][F^-]}{[HF]} = 7.2 \times 10^{-4}$$

4. The initial concentrations are

   $[HF]_0 = 1.00$ *M* (assume no dissociation has yet occurred)

   $[F^-]_0 = 0.50$ *M* (from the dissolved NaF)

   $[H^+]_0 = 0$ (ignore $H_2O$)

5. Let $x$ be the number of mol/L of HF that dissociate to reach equilibrium, producing $x$ mol/L $H^+$ and $x$ mol/L $F^-$.

6. The equilibrium concentrations are

$$[HF] = 1.00 - x$$
$$[F^-] = 0.50 + x$$
$$[H^+] = x$$

7. Now put these quantities into the $K_a$ expression for HF:

$$7.2 \times 10^{-4} = K_a = \frac{[H^+][F^-]}{[HF]} = \frac{(x)(0.50 + x)}{1.00 - x}$$

8. Simplify the expression by neglecting $x$ where possible. The terms

$$1.00 - x \quad \text{and} \quad 0.50 + x$$

occur in step 7. Assume that $x$ can be neglected in both terms. This gives the expression

$$K_a = 7.2 \times 10^{-4} \approx \frac{(x)(0.50)}{1.00}$$

Solving this expression for $x$ gives

$$x = 1.4 \times 10^{-3}$$

9. Check the assumption in step 8:

$$\frac{1.4 \times 10^{-3}}{1.00} \times 100 = 1.4\%$$

$$\frac{1.4 \times 10^{-3}}{0.50} \times 100 = 2.8\%$$

In both cases, $x$ can be neglected, since the "5% rule" is obeyed.

10. Calculate $[H^+]$.

From the definition of $[H^+]$ in step 6,

$$[H^+] = x = 1.4 \times 10^{-3}\ M$$
$$pH = -\log[H^+] = -\log(1.4 \times 10^{-3}) = 2.85$$

Compare the $[H^+]$ from sample exercise 2.6 with that obtained previously for 1.0 *M* HF (sample exercise 2.4):

| *1.0 M HF* | *1.0 M HF, 0.50 M NaF* |
|---|---|
| $[H^+] = 2.7 \times 10^{-2}\ M$ | $[H^+] = 1.4 \times 10^{-3}\ M$ |
| $pH = 1.57$ | $pH = 2.85$ |

In the solution containing dissolved NaF, the $[H^+]$ is much smaller. This is exactly the result expected from Le Châtelier's Principle. The equilibrium is

$$HF(aq) \rightleftarrows H^+(aq) + F^-(aq)$$

and the presence of extra $F^-$ should drive the position of the equilibrium to the left, lowering the $[H^+]$. This is the observed result.

This effect is sometimes called the *common ion effect:* When a solution contains a salt having an ion common with one in the equilibrium, the position of the equilibrium is driven away from the side containing that ion.

**TEST 2.5** A solution contains 0.050 $M$ HA and 0.025 $M$ NaA ($Na^+$, $A^-$) and has a pH of 4.0. Calculate the $K_a$ value for HA. Remember to use the steps shown in sample exercise 2.6.

## 2.6 Calculating the pH for Solutions of Bases

In the Brønsted-Lowry model, a base is defined as a proton acceptor. When a base dissolves in water, it reacts with some of the molecules of water, pulling off protons:

$$B(aq) + H_2O(l) \rightleftarrows BH^+(aq) + OH^-(aq)$$

Here "B" is used as a general symbol for a base other than $OH^-$. Note that the B molecules are competing with $OH^-$ for protons.

It is useful to think of bases in two categories: (1) those that contain $OH^-$, and (2) those, like B above, that do not contain $OH^-$ but generate it by reaction with water.

### Bases That Contain $OH^-$

Typically, these bases are soluble metal hydroxides such as NaOH($s$), KOH($s$), and $Ca(OH)_2(s)$. When these compounds dissolve in water, they completely dissociate to produce metal cations and hydroxide ions. Consider, for example, NaOH($s$):

$$NaOH(s) \xrightarrow{H_2O} Na^+(aq) + OH^-(aq)$$

By analogy with completely dissociated acids, which are called "strong acids," these bases are called "strong bases."

**Sample Exercise 2.7** For a 1.0 $M$ NaOH solution calculate

**A.** $[OH^-]$
**B.** $[H^+]$
**C.** pH

**Solution**

**A.** Since the NaOH is completely dissociated in solution, the solution contains $H_2O$, $Na^+$ (1.0 $M$), and $OH^-$ (1.0 $M$). The $[OH^-]$ produced from autoionization of $H_2O$ is very small and can be ignored. Therefore

$$[OH^-] = 1.0\ M$$

**B.** What is the $[H^+]$?

To find the $[H^+]$, remember that water is in equilibrium with $H^+(aq)$ and $OH^-(aq)$:

$$H_2O(l) \rightleftarrows H^+(aq) + OH^-(aq) \qquad K_w = 1.0 \times 10^{-14} = [H^+][OH^-]$$

In part A, $[OH^-]$ was determined to be 1.0 $M$, so

$$[H^+][OH^-] = [H^+](1.0\ M) = 1.0 \times 10^{-14}$$

Thus

$$[H^+] = 1.0 \times 10^{-14}$$

**C.** What is the pH?

$$pH = -\log(1.0 \times 10^{-14}) = -(-14.00) = 14.00$$

This is a basic solution: The $[OH^-]$ is greater than the $[H^+]$. The pH is greater than 7.

**TEST 2.6** Calculate the pH of a 0.10 *M* KOH solution by answering the following questions to solve the problem:

**A.** What are the major species present in the solution?
**B.** Which component will determine the $[OH^-]$?
**C.** Calculate the $[OH^-]$.
**D.** Calculate the $[H^+]$.
**E.** Calculate the pH.

## Bases That Do Not Contain $OH^-$

These are typically molecules, such as $NH_3$, which have at least one pair of unshared electrons. This pair of electrons competes with a pair of electrons on the hydroxide ion for the proton:

$$H_2\ddot{N}H + H{-}\ddot{O}{-}H \rightarrow H{-}N^+H_3 + (:\ddot{O}{-}H)^-$$

or, in general,

$$B(aq) + H_2O(l) \rightleftarrows BH^+(aq) + OH^-(aq)$$

The equilibrium constant expression for this reaction is the definition of the **base dissociation constant, $K_b$**:

$$K_b = \frac{[BH^+][OH^-]}{[B]}$$

where $[H_2O(l)]$ is left out as usual.

Note the definition of $K_b$. It *always* refers to a reaction in which a base reacts with water to produce $OH^-$ and the conjugate acid.

Bases of this type are often called **weak bases**. Since B is competing with $OH^-$ for $H^+$ in this reaction, it is accurate to describe B as a weak base compared to $OH^-$.

**Sample Exercise 2.8** Calculate the $[OH^-]$ and the pH for a 15.0 *M* $NH_3$ solution. The $K_b$ for $NH_3$ is $1.8 \times 10^{-5}$.

**Solution**

1. First, list the major species in solution: $H_2O$, $NH_3$.
   (a) $H_2O$ is both an acid and a base, and it produces small concentrations of both $H^+$ and $OH^-$:

$$H_2O(l) \rightleftarrows H^+(aq) + OH^-(aq) \qquad K_w = 10^{-14}$$

   (b) Since the $K_b$ is given for $NH_3$, it must be a base. It will therefore react with $H_2O$:

$$NH_3(aq) + H_2O(l) \rightleftarrows NH_4^+(aq) + OH^-(aq) \qquad K_b = 1.8 \times 10^{-5}$$

2. Select the dominant equilibrium.

   Since the $K_b$ is much larger than $K_w$, nearly all the $OH^-$ in this solution at equilibrium results from the reaction of $NH_3$ with $H_2O$.

3. Write the equilibrium expression for the dominant equilibrium:

$$K_b = \frac{[NH_4^+][OH^-]}{[NH_3]} = 1.8 \times 10^{-5}$$

4. List the initial concentrations:

$$[NH_3]_0 = 15.0\ M \quad \text{(no } NH_3 \text{ has yet reacted)}$$
$$[NH_4^+]_0 = 0$$
$$[OH^-]_0 = 10^{-7} \approx 0 \quad \text{(assume the contribution from the autoionization of } H_2O \text{ is negligible)}$$

5. Define $x$:

   $x$ mol/L of $NH_3$ react with water to reach equilibrium.

6. Write the equilibrium concentrations of relevant species:

$$[NH_3] = [NH_3]_0 - x = 15.0 - x$$
$$[NH_4^+] = [NH_4^+]_0 + x = 0 + x = x$$
$$[OH^-] = [OH^-]_0 + x = 0 + x = x$$

7. Substitute into the equilibrium expression:

$$K_b = 1.8 \times 10^{-5} = \frac{[NH_4^+][OH^-]}{[NH_3]} = \frac{(x)(x)}{15.0 - x}$$

8. Simplify the expression by neglecting $x$ where possible; then solve for $x$:

   Assume $15.0 - x \approx 15.0$

$$K_b = 1.8 \times 10^{-5} \approx \frac{x^2}{15.0}$$

$$x \approx 1.6 \times 10^{-2}\ M$$

9. Check the assumption in step 8:

$$\frac{x}{[NH_3]_0} \times 100 = \frac{1.6 \times 10^{-2}}{15.0} \times 100 = 0.11\%$$

   Thus the assumption that $15.0 - x = 15.0$ is valid, and the value of $x$ calculated in step 8 is correct.

10. Calculate $[H^+]$.

    From step 8, $[OH^-] = 1.6 \times 10^{-2}\ M$. Since we know that $K_w = [H^+][OH^-] = 1.0 \times 10^{-14}$ for $H_2O$, we can write

$$[H^+] = \frac{1.0 \times 10^{-14}}{[OH^-]} = \frac{1.0 \times 10^{-14}}{1.6 \times 10^{-2}} = 6.3 \times 10^{-13}\ M$$

11. Calculate the pH:

$$pH = -\log[H^+] = -\log(6.3 \times 10^{-13})$$
$$= 12.20$$

    Note that the $[NH_3]$ at equilibrium in 15.0 $M$ $NH_3$ is

$$15.0 - 1.6 \times 10^{-2} = 15.0\ M$$

to the correct number of significant figures. When $NH_3$ is dissolved in water, only a small percentage of the molecules react. In 15.0 $M$ $NH_3$:

$$\% \, NH_3 \text{ reacting} = \frac{[NH_4^+]}{[NH_3]_0} \times 100 = \frac{1.6 \times 10^{-2}}{15.0} \times 100 = 0.11\%$$

Often a 15.0 $M$ aqueous solution of $NH_3$ is labeled 15.0 $M$ $NH_4OH$. From the above calculations, it can be seen that the latter is not a very accurate description of this solution, since it contains only small amounts of $NH_4^+$ and $OH^-$.

**TEST 2.7**

**A.** To what reaction does $K_b$ always refer?

**B.** Define $K_b$ in words.

**C.** Does $K_b$ for $NH_3$ equal $1/K_a$ for $NH_4^+$? Prove your answer.

**D.** Consider a solution containing 1.0 $M$ methylamine ($CH_3NH_2$), which has a $K_b$ value of $4.38 \times 10^{-4}$.

1. List the major species in the solution.
2. Write the equation for the reaction of $CH_3NH_2$ with $H_2O$. Hint: Remember the definition of $K_b$. Even though $CH_3NH_2$ may be unfamiliar, you should be able to write the $K_b$ reaction.
3. Will the autoionization of water or the reaction of $CH_3NH_2$ with water dominate in the production of $OH^-$?
4. Write the $K_b$ equilibrium expression for $CH_3NH_2$.
5. List the initial concentrations for the species in the dominant equilibrium.
6. Define $x$.
7. Write the equilibrium concentrations in terms of $x$.
8. Substitute the equilibrium concentrations into the equilibrium expression and solve for $x$. Make the usual assumption and check its validity.
9. Calculate the pH.

**E.** Calculate the pH of a 0.10 $M$ solution of pyridine ($C_5H_5N$), which has a $K_b$ value of $1.4 \times 10^{-9}$. Be sure to use the usual steps.

## 2.7 Buffered Solutions

A **buffered solution** is a solution that resists a change in its pH when $H^+$ or $OH^-$ is added. A very important buffered solution is blood: it can absorb the acids and bases produced in biological reactions without significantly changing pH. This is very important, because cells can survive only in a very narrow pH range.

A solution can be buffered at acidic or basic pH. To buffer a solution at an acidic pH, a weak acid (HA) and its salt (usually a sodium salt, NaA, because of its high solubility) are used. To buffer a solution at a basic pH, a weak base (B) and its salt (such as BHCl, which contains $BH^+$ and $Cl^-$ ions) are used. The way in which these solutions resist a pH change will be discussed in this section.

## Acidic Buffers

**Sample Exercise 2.9** To explore the properties of acidic buffers, consider a solution that contains 0.10 $M$ acetic acid, $HC_2H_3O_2$, often abbreviated as HOAc ($K_a = 1.8 \times 10^{-5}$), and 0.10 $M$ sodium acetate, $NaC_2H_3O_2$ (abbreviated as NaOAc).

**A.** What is the pH of this solution? (This is identical to the type of problems solved in section 2.5.)

**Solution**

1. Remember, the first thing that should *always* be done is to write down the major species in the solution, in this case, HOAc, $Na^+$, $OAc^-$, $H_2O$. Remember that when a salt, such as NaOAc, dissolves, it dissociates into its ions.
2. What equilibrium will be dominant in this solution? To answer this question, go through the list of species:
(a) HOAc — Weak acid, $K_a = 1.8 \times 10^{-5}$
(b) $Na^+$ — Neither an acid nor a base
(c) $OAc^-$ — Conjugate base of HOAc
(d) $H_2O$ — Weaker acid than HOAc ($K_w$ is much less than $K_a$)
Note that the HOAc dissociation reaction

$$\text{HOAc}(aq) \rightleftarrows \text{H}^+(aq) + \text{OAc}^-(aq)$$

involves both HOAc and $OAc^-$ and is the dominant equilibrium.
3. Following normal procedures for a weak acid problem, write the equilibrium expression for the dominant equilibrium

$$K_a = \frac{[\text{H}^+][\text{OAc}^-]}{[\text{HOAc}]} = 1.8 \times 10^{-5}$$

4–6.

| *Initial Concentrations* | | *Equilibrium Concentrations* |
|---|---|---|
| $[\text{HOAc}]_0 = 0.10M$ | Let $x$ mol/L HOAc dissociate → | $[\text{HOAc}] = 0.10 - x$ |
| $[\text{OAc}^-]_0 = 0.10\ M$ | | $[\text{OAc}^-] = 0.10 + x$ |
| (why is $[\text{OAc}^-]_0 \neq 0$?) | | |
| $[\text{H}^+]_0 \approx 0$ | | $[\text{H}^+] = x$ |

Note that the three steps (initial concentrations, definition of $x$, and the equilibrium concentrations) have been represented in a condensed form.

7,8. Now substitute into $K_a$, simplify, and solve for $x$:

$$1.8 \times 10^{-5} = \frac{[\text{H}^+][\text{OAc}^-]}{[\text{HOAc}]} = \frac{(x)(0.10 + x)}{(0.10 - x)} \approx \frac{(x)(0.10)}{0.10}$$

$$x = 1.8 \times 10^{-5}\ M$$

9. Checking the assumptions shows them to be valid (they pass the 5% rule), so we go onward.
10. $[\text{H}^+] = x = 1.8 \times 10^{-5}\ M$
11. $\text{pH} = -\log[\text{H}^+] = 4.74$

**B.** What happens to the pH of this solution if 0.010 mol of NaOH($s$) is added to 1.0 L of this buffered solution?

**Solution**

First, write down the major species in the solution (before any reaction occurs): HOAc, $OAc^-$, $Na^+$, $OH^-$, $H_2O$.

Note that the solution contains relatively large quantities of $OH^-$, which is a strong base. It will be looking for the best source of $H^+$ with which to react. The weak acid, HOAc, can furnish $H^+$ according to the following reaction:

$$HOAc(aq) + OH^-(aq) \longrightarrow H_2O(l) + OAc^-(aq)$$

*Assume this reaction goes to completion.* ($OH^-$ is a much stronger base than $OAc^-$.) In dealing with a solution where a reaction goes to completion, it is always best to consider the reaction first, then do the equilibrium problem. A situation like this really involves two problems: (1) a stoichiometry problem, and (2) an equilibrium problem. First consider the stoichiometry problem involved when $OH^-$ reacts to completion with HOAc.

1. The stoichiometry problem.
   0.010 mol of NaOH(*s*) has been added to 1.0 L of a solution containing 0.10 *M* HOAc and 0.10 *M* NaOAc. The reaction is

| | $OH^-$ + | HOAc | → $H_2O$ + | $OAc^-$ |
|---|---|---|---|---|
| Before the reaction | 0.01 mol | (1.0 L × 0.10 *M*) = 0.10 mol | | 1.0 L × 0.10 *M* = 0.10 mol |
| After the reaction | 0 | 0.10 − 0.01 = 0.09 mol | | 0.10 + 0.01 = 0.11 mol |

2. The equilibrium problem.
   After the reaction, the solution contains $Na^+$, $OAc^-$, HOAc, $H_2O$. As before, use the weak acid dissociation reaction:

$$HOAc(aq) \rightleftarrows H^+(aq) + OAc^-(aq)$$

| *Initial Concentrations* | | *Equilibrium Concentrations* |
|---|---|---|
| $[HOAc]_0 = \frac{0.09 \text{ mol}}{1.0 \text{ L}} = 0.09\ M$ | Let *x* mol/L of HOAc dissociate → | $[HOAc] = 0.09 - x$ |
| $[OAc^-]_0 = \frac{0.11 \text{ mol}}{1.0 \text{ L}} = 0.11\ M$ | | $[OAc^-] = 0.11 + x$ |
| $[H^+]_0 = 0$ | | $[H^+] = x$ |

$$1.8 \times 10^{-5} = K_a = \frac{[H^+][OAc^-]}{[HOAc]} = \frac{(x)(0.11 + x)}{(0.09 - x)} \approx \frac{(x)(0.11)}{0.09}$$

$$x \approx \frac{0.09}{0.11}(1.8 \times 10^{-5}) = 1.5 \times 10^{-5}\ M$$

Note that the assumptions are valid. Thus

$$[H^+] = 1.5 \times 10^{-5}\ M$$

$$pH = -\log(1.5 \times 10^{-5}) = 4.82$$

The pH of the solution (containing 0.10 *M* HOAc and 0.10 *M* NaOAc) before addition of NaOH was 4.74. After the addition of 0.01 mol of NaOH(*s*), the pH is 4.82. The increase is only 0.08 pH unit. Contrast this pH change with the one that occurs when 0.010 mol of NaOH(*s*) is added to 1.0 L of water; in that case, the pH changes from 7 to 12, as we will show.

1. Pure water has pH = 7.0.
2. Addition of 0.01 mol NaOH(*s*) to 1.0 L of water produces a 0.01 *M* solution of NaOH. This is a strong base (completely dissociated), so that

$$[OH^-] = 0.010\ M$$

$$1.0 \times 10^{-14} = K_w = [H^+][OH^-] = [H^+](0.010\ M)$$

$$[H^+] = \frac{1.0 \times 10^{-14}}{1.0 \times 10^{-2}} = 1.0 \times 10^{-12}$$

$$pH = -\log(1.0 \times 10^{-12}) = 12.00$$

The pH increases from 7.0 (pure water) to 12.0 (0.01 *M* NaOH solution), which is an increase of 5 pH units. Thus, adding a given quantity of $OH^-$ to 1.0 L of water produces a much greater change in pH than adding the same quantity of $OH^-$ to 1.0 L of a buffered solution. Why is this true? How is buffering accomplished?

Remember that an acidic buffer contains large quantities of HA and $A^-$. The pH is governed by the dissociation equilibrium

$$HA(aq) \rightleftarrows H^+(aq) + A^-(aq) \qquad K_a = \frac{[H^+][A^-]}{[HA]}$$

Rearranging the $K_a$ expression,

$$[H^+] = K_a \frac{[HA]}{[A^-]}$$

Notice from this form of the $K_a$ expression that the $[H^+]$ depends on the $[HA]/[A^-]$ ratio. The $[H^+]$ changes to the extent that this ratio changes.

When $OH^-$ is added to a buffered solution, the reaction

$$OH^- + HA \longrightarrow A^- + H_2O$$

occurs; i.e., HA is changed to $A^-$. Now if [HA] and $[A^-]$ are both large compared to the amount of $OH^-$ added, the percentage change in the $[HA]/[A^-]$ ratio will be small, and thus the change in $[H^+]$ will be small.

On the other hand, if either [HA] or $[A^-]$ is small, a much larger percentage change in $[HA]/[A^-]$ will occur and $[H^+]$ will change more significantly. Thus, buffers are best prepared by using nearly equal amounts of HA and $A^-$.

When a strong acid is added to a buffered solution containing HA and $A^-$, the reaction

$$H^+ + A^- \longrightarrow HA$$

can be assumed to go to completion. In this case, $A^-$ is changed to HA.

**TEST 2.8** Consider a solution containing 0.20 *M* HOAc ($K_a = 1.8 \times 10^{-5}$) and 0.10 *M* NaOAc.

**A.** Calculate the pH of this solution.
**B.** Calculate the pH of the solution after 0.020 mol of HCl(*g*) is dissolved in 1.0 L of this solution.
**C.** How much does the pH of the original buffer change upon addition of 0.020 mol of NaOH(*s*) to 1.0 L of this solution?

## Choosing a Buffering System

A buffer is effective in maintaining a particular pH only to the extent that the ratio $[HA]/[A^-]$ is insensitive to added $H^+$ or $OH^-$. The ratio $[HA]/[A^-]$ is most insensitive to change when $[HA] = [A^-]$. This is an important point to remember in choosing the buffering system.

**Sample Exercise 2.10** A chemist who wishes to prepare a solution buffered at pH = 4.0 can choose among the following acids:

| *Weak Acid* | $K_a$ |
|---|---|
| HA | $1.0 \times 10^{-2}$ |
| HB | $1.0 \times 10^{-4}$ |
| HC | $1.0 \times 10^{-6}$ |

Which acid (along with its conjugate base) should the chemist choose?

**Solution**

The solution is to be buffered at pH = 4.0. This means that $[H^+] = 1.0 \times 10^{-4}\ M$. Rearrange the $K_a$ expression:

$$[H^+] = K_a \frac{[\text{acid}]}{[\text{anion}]} = 1.0 \times 10^{-4}$$

The most effective buffering will occur when [acid] / [anion] = 1.0. Thus

$$[H^+] = 1.0 \times 10^{-4} = K_a(1.0)$$

Therefore the acid (HB) with $K_a = 1.0 \times 10^{-4}$ should be chosen, and a comparable amount of $B^-$ added to make the buffer.

## The Capacity of a Buffer

The **capacity** of a buffer reflects its ability to absorb $H^+$ or $OH^-$ without significantly changing pH. The more buffering material present in the solution, the greater the capacity of the buffer.

**TEST 2.9** Consider two solutions:

**I.** 1.0 *M* HA ($K_a = 1.0 \times 10^{-5}$), 1.0 *M* NaA

**II.** 0.10 *M* HA ($K_a = 1.0 \times 10^{-5}$), 0.10 *M* NaA

**A.** Calculate the pH of each solution.
**B.** Which buffered solution has the greater capacity?
**C.** Support your answer by calculating the change in pH that occurs when 0.050 mol of $OH^-$ is added to 1.0 L of each solution.

## Basic Buffers

To buffer a solution at a basic pH requires a weak base and its salt. For example, one might use $NH_3$ ($K_b = 1.8 \times 10^{-5}$) and $NH_4Cl$.

**Sample Exercise 2.11** Consider a solution containing 0.30 *M* $NH_3$ and 0.20 *M* $NH_4Cl$.

**A.** Calculate the pH of this solution.

**Solution**

The solution contains the major species $NH_3$, $NH_4^+$, $Cl^-$, $H_2O$.

1. $NH_3$ is a base:

$$NH_3(aq) + H_2O(l) \rightleftarrows NH_4^+(aq) + OH^-(aq)$$

2. $NH_4^+$ is the conjugate acid of $NH_3$.
3. $Cl^-$ is a very weak base (it is the anion of HCl) and will not affect the pH.
4. $H_2O$ is a weak acid and base:

$$H_2O(l) \rightleftarrows H^+(aq) + OH^-(aq)$$

The dominant equilibrium involves $NH_3$ and $NH_4^+$:

$$NH_3(aq) + H_2O(l) \rightleftarrows NH_4^+(aq) + OH^-(aq)$$

$$K_b = \frac{[NH_4^+][OH^-]}{[NH_3]} = 1.8 \times 10^{-5}$$

| *Initial Concentrations* | | *Equilibrium Concentrations* |
|---|---|---|
| $[NH_3]_0 = 0.30\ M$ | Let $x$ mol/L $NH_3$ react with $H_2O$ → | $[NH_3] = 0.30 - x$ |
| $[NH_4^+]_0 = 0.20\ M$ | | $[NH_4^+] = 0.20 + x$ |
| $[OH^-]_0 \approx 0$ (ignore water) | | $[OH^-] = x$ |

$$K_b = 1.8 \times 10^{-5} = \frac{[NH_4^+][OH^-]}{[NH_3]} = \frac{(0.20 + x)(x)}{0.30 - x}$$

Neglect $x$ compared to 0.30 and 0.20:

$$1.8 \times 10^{-5} \approx \frac{(0.20)(x)}{0.30}$$

$$x \approx 2.7 \times 10^{-5}$$

Application of the "5% rule" shows the approximations to be valid. Thus

$$[OH^-] = x = 2.7 \times 10^{-5}$$

From the water equilibrium

$$[H^+] = \frac{K_w}{[OH^-]} = \frac{1.0 \times 10^{-14}}{2.7 \times 10^{-5}} = 3.7 \times 10^{-10}$$

$$pH = 9.43$$

**B.** Calculate the pH after 0.050 mol of HCl is added to this solution.

**Solution**

After the HCl has been added, but before any reaction occurs, the solution contains $H^+$, $Cl^-$, $NH_4^+$, $NH_3$, $H_2O$. Note that the solution contains large quantities of $H^+$ and $NH_3$. The reaction between them will go essentially to completion.

Do the stoichiometry problem first.

| | $H^+$ | + | $NH_3$ | → | $NH_4^+$ |
|---|---|---|---|---|---|
| Before the reaction | 0.050 mol | | 0.30 mol | | 0.20 mol |
| After the reaction | 0 | | 0.30 − 0.050 = 0.25 mol | | 0.20 + 0.050 = 0.25 mol |

Now do the equilibrium problem. After the reaction, the major species are

$$NH_4^+, Cl^-, NH_3, H_2O$$

The dominant equilibrium is

$$NH_3(aq) + H_2O(l) \rightleftarrows NH_4^+(aq) + OH^-(aq) \qquad K_b = \frac{[NH_4^+][OH^-]}{[NH_3]}$$

| *Initial Concentrations* | | *Equilibrium Concentrations* |
|---|---|---|
| $[NH_3]_0 = 0.25\ M$ | Let $x$ mol/L $NH_3$ react with $H_2O$ $\longrightarrow$ | $[NH_3] = 0.25 - x$ |
| $[NH_4^+]_0 = 0.25\ M$ | | $[NH_4^+] = 0.25 + x$ |
| $[OH^-]_0 \approx 0$ (ignore $H_2O$) | | $[OH^-] = x$ |

$$K_b = 1.8 \times 10^{-5} = \frac{(0.25 + x)(x)}{0.25 - x} \approx \frac{(0.25)(x)}{0.25}$$

$$x \approx 1.8 \times 10^{-5}$$

The "5% rule" shows the assumptions to be valid. Thus

$$[OH^-] = x = 1.8 \times 10^{-5}\ M$$

From the $H_2O$ equilibrium,

$$[H^+] = \frac{K_w}{[OH^-]} = \frac{1.0 \times 10^{-14}}{1.8 \times 10^{-5}} = 5.6 \times 10^{-10}\ M$$

$$pH = 9.25$$

**C.** Calculate the pH after 0.030 mol of $NaOH(s)$ is added to the original solution.

**Solution**

After the NaOH is added, but before any reaction occurs, the solution contains $Na^+$, $OH^-$, $NH_4^+$, $NH_3$, $H_2O$. The $OH^-$ is a very strong base and will react essentially completely with the $NH_4^+$.

1. The stoichiometry problem.

   Assume that the reaction goes to completion:

| | $OH^-$ | + | $NH_4^+$ | $\longrightarrow$ | $NH_3$ | + | $H_2O$ |
|---|---|---|---|---|---|---|---|
| Before the reaction | 0.030 mol | | 0.20 mol | | 0.30 mol | | |
| After the reaction | 0 | | 0.20 − 0.03 = 0.17 mol | | 0.30 + 0.03 = 0.33 mol | | |

2. The equilibrium problem.

$$K_b = \frac{[NH_4^+][OH^-]}{[NH_3]}$$

| *Initial Concentrations* | | *Equilibrium Concentrations* |
|---|---|---|
| $[NH_3]_0 = 0.33\ M$ | Let $x$ mol/L $NH_3$ react with $H_2O$ $\longrightarrow$ | $[NH_3] = 0.33 - x$ |
| $[NH_4^+]_0 = 0.17\ M$ | | $[NH_4^+] = 0.17 + x$ |
| $[OH^-]_0 \approx 0$ (ignore $H_2O$) | | $[OH^-] = x$ |

$$K_b = 1.8 \times 10^{-5} = \frac{(0.17 + x)(x)}{0.33 - x} \approx \frac{(0.17)(x)}{0.33}$$

$$x \approx 3.5 \times 10^{-5}$$

The "5% rule" shows the approximations to be valid. Thus

$$[OH^-] = x = 3.5 \times 10^{-5}\ M$$

From the $H_2O$ equilibrium,

$$[H^+] = \frac{K_w}{[OH^-]} = \frac{1.0 \times 10^{-14}}{3.5 \times 10^{-5}} = 2.9 \times 10^{-10}$$

$$pH = 9.54$$

**TEST 2.10**

**A.** Consider a solution containing 0.50 *M* HA ($K_a = 1.0 \times 10^{-6}$) and 0.50 *M* NaA.
1. Calculate the pH of this solution.
2. Calculate the pH of the solution after 0.10 mol of NaOH(*s*) is added to 1.0 L of this solution.
3. Calculate the pH of the solution after 0.10 mol of HCl(*g*) is added to 1.0 L of this solution.

**B.** Consider the following two solutions:

**I.** 0.50 *M* HF ($K_a = 7.2 \times 10^{-4}$) and 0.50 *M* NaF

**II.** 5.0 *M* HF and 5.0 *M* NaF

1. Calculate the pH of each solution.
2. Which buffered solution has the largest capacity?
3. Support your answer by calculating the change in pH when 0.05 mol of HCl(*g*) is added to 1.0 L of each solution.

**C.** What concentration of $NaNO_2$ is necessary to buffer 0.050 *M* $HNO_2$ ($K_a = 4.0 \times 10^{-4}$) at pH = 3.00?

**D.** What is the pH of a solution formed by mixing 50.0 mL of 0.050 *M* HCl with 100.0 mL of 0.10 *M* NaOAc? The $K_a$ for HOAc is $1.8 \times 10^{-5}$.

**E.** A solution is prepared by adding 0.100 mol of NaOH(*s*) to 1.0 L of 0.500 *M* $NH_4Cl$. Calculate the pH. The $K_b$ for $NH_3$ is $1.8 \times 10^{-5}$.

## 2.8 Acid-Base Reactions That Occur When Salts Are Dissolved in Water*

"Salt" is simply another name for an ionic compound. When a typical salt is dissolved in water, it breaks up into its ions, which, in dilute solution, move about independently. Some salts, when dissolved in water, form an acidic solution; others form a basic solution; and still others are neutral in water.

Ions may behave as weak acids. For example, when $NH_4Cl$ is dissolved in water, $NH_4^+$ and $Cl^-$ are produced. The $NH_4^+$ is a weak acid:

$$H_2O(l) + NH_4^+(aq) \rightleftarrows NH_3(aq) + H_3O^+(aq)$$

Here $NH_3$ and $H_2O$ compete for the proton.

Note that $NH_4^+$ is the conjugate acid of the base, $NH_3$. There are many similar salts in which the *cation is the conjugate acid of a base;* such salts produce acidic solutions when dissolved in water.

A second kind of salt involves ions that produce an acidic solution but that are not in themselves Brønsted-Lowry acids. These ions gain their acidic character from

*Called hydrolysis in most textbooks.

waters of hydration. All ions are hydrated when they are placed in water. In many cases, these ions have no effect on the acidity of the water molecules. However, when *small, highly charged ions* are hydrated, the tendency for the hydrating water molecules to give up $H^+$ is much greater than for unattached water molecules. For example, when $AlCl_3$ is dissolved in water,

$$AlCl_3(s) \xrightarrow{H_2O} Al^{3+}(aq) + 3Cl^-(aq)$$

the $Al^{3+}$ ion is hydrated by six water molecules to form $Al(OH_2)_6^{3+}$:

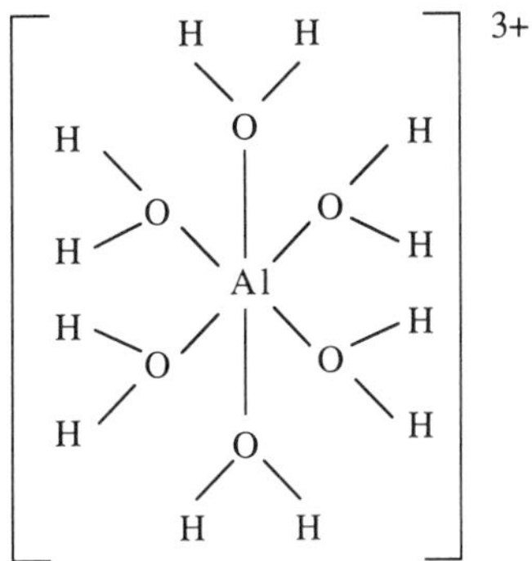

This hydrated ion undergoes the reaction

$$Al(OH_2)_6^{3+}(aq) + H_2O(l) \rightleftharpoons Al(OH)(OH_2)_5^{2+}(aq) + H_3O^+(aq)$$

The charge on the metal ion causes the attached water molecules to be more acidic than free $H_2O$.

A third case involves salts in which the anion behaves as a base. For example, a solution of NaCN contains $Na^+$ and $CN^-$ ions. The $CN^-$ ion has a high affinity for $H^+$ (we know this because HCN, $K_a = 6.2 \times 10^{-10}$, is a very weak acid) and thus produces a basic solution via the reaction

$$CN^-(aq) + H_2O(l) \rightleftharpoons HCN(aq) + OH^-(aq)$$

*This behavior is characteristic of the anion of any weak acid.*

In summary, there are these important cases to consider:

1. Acid solutions result when the following are dissolved:
   (a) Salts that contain the conjugate acid of a base.
   (b) Salts that contain small, highly charged cations.
2. A basic solution results when a salt containing the anion of a weak acid is dissolved. Each of these will now be considered in more detail.

## Acidic Solutions: Case 1

First we consider salts containing the conjugate acid of a base. To illustrate this case, consider $NH_4Cl$ dissolved in water to produce $NH_4^+$ and $Cl^-$. The $NH_4^+$ behaves as a weak acid:

$$NH_4^+(aq) + H_2O(l) \rightleftharpoons NH_3(aq) + H_3O^+(aq)$$

which is more commonly represented as a simple dissociation:

$$NH_4^+(aq) \rightleftharpoons NH_3(aq) + H^+(aq) \qquad K_a = \frac{[NH_3][H^+]}{[NH_4^+]}$$

Solutions of $NH_4^+$ can be treated in the same way as those of any weak acid. The only difference is that the $K_a$ for $NH_4^+$ is not usually listed in tables of acid-base equilibrium constants. Rather, $K_b$ for $NH_3$ is given.

How can the $K_b$ for $NH_3$ be used to calculate the $K_a$ for $NH_4^+$? Remember the following important definitions:

$$K_b(NH_3) = \frac{[NH_4^+][OH^-]}{[NH_3]} = 1.8 \times 10^{-5}$$

$$K_a(NH_4^+) = \frac{[NH_3][H^+]}{[NH_4^+]} = ?$$

$$K_w(H_2O) = [H^+][OH^-] = 1.00 \times 10^{-14}$$

Note that

$$\underbrace{\frac{[\cancel{NH_3}][H^+]}{[\cancel{NH_4^+}]}}_{K_a(NH_4^+)} \times \underbrace{\frac{[\cancel{NH_4^+}][OH^-]}{[\cancel{NH_3}]}}_{K_b(NH_3)} = \underbrace{[H^+][OH^-]}_{K_w(H_2O)}$$

Thus

$$K_a \times K_b = K_w$$

This is true for

1. Any base, B, and its conjugate acid, $BH^+$.
2. Any acid, HX, and its conjugate base, $X^-$.

This relationship is very useful, as can be seen in the sample exercises in this section.

**Sample Exercise 2.12** Calculate the pH of a solution prepared by dissolving 0.010 mol of $NH_4Cl(s)$ in enough water to make 1.0 L of solution. The $K_b$ for $NH_3$ is $1.8 \times 10^{-5}$.

**Solution**

To solve this problem, recognize that $NH_4Cl(s)$ is an ionic solid, containing $NH_4^+$ and $Cl^-$, which breaks up into its ions when it dissolves:

$$NH_4Cl(s) \xrightarrow{H_2O} NH_4^+(aq) + Cl^-(aq)$$

so that the major species in solution are $H_2O$, $NH_4^+$, $Cl^-$. To decide which of the reactions is dominant, consider the acid or base properties of each major component:

1. $H_2O$ Acid or base (very weak)
2. $NH_4^+$ Weak acid (it can lose a proton to form $NH_3$)
3. $Cl^-$ Extremely weak base (we know this because HCl is a very strong acid)

Will $H_2O$ or $NH_4^+$ be the dominant acid?

To answer this question, find the acid dissociation constant for each. $K_w$ is $10^{-14}$. What is $K_a$ for $NH_4^+$? $K_b$ for $NH_3$ is given as $1.8 \times 10^{-5}$. Thus

$$K_b(NH_3) \times K_a(NH_4^+) = K_w$$

$$K_a(NH_4^+) = \frac{K_w}{K_b(NH_3)} = \frac{10^{-14}}{1.8 \times 10^{-5}} = 5.6 \times 10^{-10}$$

Thus $NH_4^+$ is a stronger acid than $H_2O$ and will dominate in the production of $H^+$, making the reaction of interest

$$NH_4^+(aq) \rightleftarrows NH_3(aq) + H^+(aq) \qquad K_a = \frac{[NH_3][H^+]}{[NH_4^+]} = 5.6 \times 10^{-10}$$

Note the following:

1. Just because $K_b$ for $NH_3$ is given in this problem does not mean that the $K_b$ reaction should be used. Let the solution components tell you what reaction to use. This solution contains $NH_4^+$, not $NH_3$. Thus $NH_4^+$ is the reactant, not $NH_3$.
2. This is a typical weak acid problem. The only differences between this case and previous weak acid problems are that the acid in this problem is ionic and the $K_a$ for the acid is not given, but must be calculated from $K_b$ and $K_w$.

The problem may now be solved in the normal way:

| *Initial Concentrations* | | *Equilibrium Concentrations* |
|---|---|---|
| $[NH_4^+]_0 = 0.010\ M$ | Let $x$ mol/L of $NH_4^+$ dissociate to come to equilibrium $\longrightarrow$ | $[NH_4^+] = 0.010 - x$ |
| $[H^+]_0 \approx 0$ | | $[H^+] = x$ |
| $[NH_3]_0 = 0$ | | $[NH_3] = x$ |

$$K_a = 5.6 \times 10^{-10} = \frac{[NH_3][H^+]}{[NH_4^+]} = \frac{(x)(x)}{0.010 - x} \approx \frac{x^2}{0.010}$$

$$x^2 = 5.6 \times 10^{-12}$$

$$x = 2.4 \times 10^{-6}$$

Noting that the assumption is valid,

$$[H^+] = x = 2.4 \times 10^{-6}$$

$$pH = 5.62$$

## Acidic Solutions: Case 2

Now we consider solutions containing highly charged cations. Recall that when small, highly charged cations are placed in water, the resulting hydrated ion may be acidic. For example, when $Co^{3+}$ is placed in water, $Co(OH_2)_6^{3+}$ is formed, which undergoes acid dissociation:

$$Co(OH_2)_6^{3+}(aq) \rightleftarrows Co(OH)\,(OH_2)_5^{2+}(aq) + H^+(aq) \qquad K_a = 1.02 \times 10^{-5}$$

**Sample Exercise 2.13** A solution is prepared by dissolving 0.050 mol of $CoCl_3(s)$ in enough water to make 1.00 L of solution. Calculate the pH of this solution. ($K_a$ for $Co(OH_2)_6^{3+}$ is $1.02 \times 10^{-5}$.)

**Solution**

Since $CoCl_3(s)$ is an ionic compound, the solution will contain the major species $Co^{3+}(aq)$, $Cl^-(aq)$, $H_2O$, where $Co^{3+}(aq)$ is $Co(OH_2)_6^{3+}$. Consider the acid-base properties of each major component:

1. $Co(OH_2)_6^{3+}$ — Weak acid ($K_a$ given)
2. $Cl^-$ — Extremely weak base (no effect on pH)
3. $H_2O$ — Acid or base (very weak)

Clearly, $Co(OH_2)_6^{3+}$ is the dominant weak acid ($K_a$ for $Co(OH_2)_6^{3+}$ is much greater than $K_w$):

$$Co(OH_2)_6^{3+}(aq) \rightleftarrows Co(OH)(OH_2)_5^{2+}(aq) + H^+(aq)$$

$$K_a = \frac{[Co(OH)(OH_2)_5^{2+}][H^+]}{[Co(OH_2)_6^{3+}]} = 1.02 \times 10^{-5}$$

Now go through the usual steps:

1. Define the initial concentrations.
2. Define $x$.
3. Write down the equilibrium concentrations.

| *Initial Concentrations* | | *Equilibrium Concentrations* |
|---|---|---|
| $[Co(OH_2)_6^{3+}]_0 = 0.050\ M$ | Let $x$ mol/L $Co(OH_2)_6^{3+}$ dissociate $\longrightarrow$ | $[Co(OH_2)_6^{3+}] = 0.050 - x$ |
| $[Co(OH)(OH_2)_5^{2+}]_0 = 0$ | | $[Co(OH)(OH_2)_5^{2+}] = x$ |
| $[H^+]_0 \approx 0$ | | $[H^+] = x$ |

$$1.02 \times 10^{-5} = K_a = \frac{[Co(OH)(OH_2)_5^{2+}][H^+]}{[Co(OH_2)_6^{3+}]} = \frac{(x)(x)}{0.050 - x} = \frac{x^2}{0.050}$$

$$x^2 \approx (0.050)(1.02 \times 10^{-5}) = 5.10 \times 10^{-7}$$

$$x \approx 7.1 \times 10^{-4}$$

Since

$$\frac{x}{[HA]_0} \times 100 = \frac{7.10 \times 10^{-4}}{5.0 \times 10^{-2}} \times 100 = 1.4\%$$

the assumption is valid, so

$$[H^+] = 7.1 \times 10^{-4}$$

$$pH = 3.15$$

Note that the solution is quite acidic.

**TEST 2.11** **A.** Calculate the pH of a 0.10 $M$ solution of $C_6H_5NH_3Cl$. The $K_b$ for $C_6H_5NH_2$ is $3.8 \times 10^{-10}$.

1. What are the major species in solution? (Hint: $C_6H_5NH_3Cl$ is a salt.)
2. What is the value of $K_a$ for $C_6H_5NH_3^+$?
3. What reaction will dominate in determining the pH?
4. Set up the equilibrium expression for (3).
5. List the initial concentrations.
6. Define $x$.
7. Write the equilibrium concentrations in terms of $x$.
8. Solve for $x$. (Check assumptions.)
9. Calculate the pH.

**B.** Calculate the pH of a 0.20 $M$ solution of $CrCl_3$. ($Cr(OH_2)_6^{3+}$ has $K_a = 1.5 \times 10^{-4}$.)

## Summary

Certain positive ions can behave as acids. In general there are two classes:

1. Positive ions that are conjugate acids of bases (that is, protonated bases).
2. Hydrated metal ions.

## Basic Solutions

When some salts are dissolved in water, a basic solution results. For example, consider what happens when sodium acetate is dissolved in water.

$$\text{NaOAc}(s) \xrightarrow{H_2O} \text{Na}^+(aq) + \text{OAc}^-(aq)$$

What are the acid or base properties of these ions?

1. $Na^+$ is neither an acid nor a base.
2. $OAc^-$ is a base. Acetate ion is the conjugate base of acetic acid (a weak acid), which means that acetate ion has a significant affinity for $H^+$.

---

**Sample Exercise 2.14** Calculate the pH of a 1.0 *M* solution of sodium acetate. The $K_a$ for acetic acid is $1.8 \times 10^{-5}$.

**Solution**

The major species are $Na^+$, $OAc^-$, $H_2O$.

1. $Na^+$ Neither an acid nor a base
2. $OAc^-$ Base (HOAc is a weak acid)
3. $H_2O$ Weak acid or base

When $OAc^-$ is placed in water, what reaction will occur? Acetate ion is a base, which means it will combine with a proton. Where will the proton come from? Water is the only source. Thus the reaction must be

$$\text{OAc}^-(aq) + H_2O(l) \rightleftarrows \text{HOAc}(aq) + \text{OH}^-(aq) \qquad K_b = \frac{[\text{HOAc}][\text{OH}^-]}{[\text{OAc}^-]}$$

What is the value of $K_b$? We know that

$$K_a\,(\text{HOAc}) \times K_b\,(\text{OAc}^-) = K_w$$

Thus

$$(1.8 \times 10^{-5})\,K_b = 1.0 \times 10^{-14}$$

$$K_b\,(\text{OAc}^-) = \frac{1.0 \times 10^{-14}}{1.8 \times 10^{-5}} = 5.6 \times 10^{-10}$$

Note that this is a very small equilibrium constant. $OAc^-$ does not compete well against $OH^-$ for $H^+$. However,

$$K_b\,(\text{OAc}^-) \gg K_w$$

so that this reaction is much more important in producing $OH^-$ than is the simple dissociation of water. Thus the dominant reaction in this solution is the $K_b$ reaction of $OAc^-$:

$$\text{OAc}^-(aq) + H_2O(l) \rightleftarrows \text{HOAc}(aq) + \text{OH}^-(aq) \qquad K_b = \frac{[\text{HOAc}][\text{OH}^-]}{[\text{OAc}^-]} = 5.6 \times 10^{-10}$$

| *Initial Concentrations* | | *Equilibrium Concentrations* |
|---|---|---|
| $[\text{OAc}^-] = 1.0\ M$ | Let $x$ mol/L $OAc^-$ react with $H_2O$ → | $[\text{OAc}^-] = 1.0 - x$ |
| $[\text{OH}^-]_0 \approx 0$ | | $[\text{OH}^-] = x$ |
| $[\text{HOAc}]_0 = 0$ | | $[\text{HOAc}] = x$ |

$$5.6 \times 10^{-10} = K_b = \frac{[OH^-][HOAc]}{[OAc^-]} = \frac{(x)(x)}{1.0 - x} \approx \frac{x^2}{1.0}$$

$$x^2 \approx (1.0)(5.6 \times 10^{-10})$$

$$x \approx 2.4 \times 10^{-5}\ M$$

Note that $1.0 \ggg 2.4 \times 10^{-5}\ M$, so the assumption is valid. Then

$$x = [OH^-] = 2.4 \times 10^{-5}\ M$$

To find the pH, use $K_w$:

$$K_w = 1.0 \times 10^{-14} = [H^+][OH^-] = [H^+](2.4 \times 10^{-5})$$

$$[H^+] = \frac{1.0 \times 10^{-14}}{2.4 \times 10^{-5}} = 4.2 \times 10^{-10}$$

$$pH = 9.38$$

Thus a 1.0 *M* sodium acetate solution is quite basic.

## Summary

When salts are dissolved in water, acid-base reactions may occur. Consider the following cases:

**A.** A basic solution will result if the anion of the salt is the conjugate base of a *weak acid.* The $K_b$ for the anion can be calculated from the $K_a$ value of the parent acid ($K_a \cdot K_b = K_w$). Anions of strong acids (e.g., $Cl^-$, $NO_3^-$, $ClO_4^-$, $HSO_4^-$) have very low affinities for $H^+$ and thus do not affect the pH. (A solution of NaCl or $KNO_3$ has pH = 7.)

**B.** An acidic solution will result when:
1. The cation is the conjugate acid of a weak base. The cation is a weak acid whose $K_a$ can be calculated from $K_b$ for the parent base ($K_a \cdot K_b = K_w$).
2. The cation is a hydrated metal ion (usually with a charge of +3 or greater), which behaves as a weak acid. The $K_a$ value for this type of weak acid will be given.

**TEST 2.12**

**A.** Consider an aqueous solution of each of the following salts. For each salt, state whether the solution will be neutral, acidic, or basic, and write the reaction for the latter two cases.
1. NaCl
2. $NH_4NO_3$
3. $Al(NO_3)_3$
4. NaCN ($K_a$ for HCN is $6.2 \times 10^{-10}$)
5. $(CH_3)_3NHCl$ ($K_b$ for $(CH_3)_3N$ is $5.3 \times 10^{-5}$)

**B.** Since acetic acid (HOAc) is a weak acid, we say that acetate ion ($OAc^-$) must be a strong base. However, the $K_b$ for $OAc^-$ is only $5.6 \times 10^{-10}$. Explain this apparent discrepancy. (Hint: compare the $K_a$ reaction for HOAc and the $K_b$ reaction for $OAc^-$. In both reactions, $OAc^-$ competes for $H^+$. Does it compete with the same base in both reactions?)

## 2.9 Henderson-Hasselbalch Equation

For the dissociation of a weak acid,

$$HA(aq) \rightleftarrows H^+(aq) + A^-(aq) \qquad K_a = \frac{[H^+][A^-]}{[HA]} = [H^+]\frac{[A^-]}{[HA]}$$

Now take the log of both sides

$$\log K_a = \log [H^+] + \log \frac{[A^-]}{[HA]}$$

(remembering that the log of a product is the sum of the logs, i.e., $\log xy = \log x + \log y$). Multiply both sides of the equation by −1:

$$-\log K_a = -\log [H^+] - \log \frac{[A^-]}{[HA]}$$

$$pK_a = pH - \log \frac{[A^-]}{[HA]}$$

$$pH = pK_a + \log \frac{[A^-]}{[HA]}$$

This "log form" of the $K_a$ expression is called the **Henderson-Hasselbalch equation.** This equation is useful in situations where the [HA] and $[A^-]$ are known.

**Sample Exercise 2.15** The acid HOCl has a $pK_a$ value of 7.50. Calculate the pH of a solution containing 0.25 *M* HOCl and 0.75 *M* NaOCl.

**Solution**

$$pH = pK_a + \log \frac{[A^-]}{[HA]}$$

$$= 7.50 + \log\left(\frac{0.75}{0.25}\right) = 7.50 + \log(3.0)$$

$$= 7.50 + 0.48 = 7.98$$

## 2.10 Review of Procedures

Remember: The most important part of doing an acid-base problem is the analysis at the beginning of the problem:

Does a reaction occur?
What is it?
What equilibrium dominates?

The best way to answer these questions successfully is to *write the major species in solution.* Then ask the question: Does a reaction occur that goes to completion? The situations to look for are:

Has $OH^-$ been added to a solution containing an acid?
Has $H^+$ been added to a solution containing a base?

In both of these situations, the reaction that occurs can be assumed to go to completion. After the reaction has been allowed to go to completion, again *write the major species.* Now check each one for acid or base properties and select the dominant equilibrium by looking at the values of the various equilibrium constants. In almost every case, one equilibrium will dominate and can be used to solve for the $[H^+]$ or $[OH^-]$.

When faced with an acid-base problem, the best strategy is to assume that it is not exactly like any other problem you may have done. One small change can cause a problem that looks very similar to one you have done before to be quite different.

When starting an acid-base problem, the **wrong** question to ask is: How can I use a problem whose solution I have memorized to solve this problem? The **correct** question is: What species are in solution and what do they do? (Think chemistry!) *Let the problem guide you.*

The steps to follow are:

1. Write the major species in solution before any reactions take place.
2. Look for any reactions taking place that can be assumed to go to completion.

   Examples: $OH^-$ with acid (strong or weak)

   $H^+$ with base (strong or weak)
3. If a reaction occurs that can be assumed complete:
   (a) Do the stoichiometry problem.
   (b) Write the major species in solution after the reaction.
4. Look at each major component of the solution and decide which are acids or bases.
5. Pick the equilibrium that will control the $[H^+]$. Use the $K$'s for the various species to help decide.
6. Do the equilibrium calculation.
   (a) Write the equation for the reaction and the equilibrium expression.
   (b) Compute the initial concentrations (assuming the equilibrium of interest has not yet occurred, i.e., no acid dissociation, etc.)
   (c) Define $x$
   (d) Compute the equilibrium concentrations in terms of $x$.
   (e) Substitute in the equilibrium expression and solve for $x$, making approximations to simplify the math if possible.
   (f) Check the validity of the approximations.
   (g) Calculate the pH.

---

## EXERCISES

1. For each of the following solutions, list the major species in solution and select the equilibrium that will control the pH.
   (a) 1.00 $M$ $HNO_2$ ($K_a = 4.0 \times 10^{-4}$)
   (b) 0.010 $M$ $HNO_3$
   (c) 1.0 $M$ $C_5H_5N$ ($K_b = 1.4 \times 10^{-9}$)
   (d) 0.10 $M$ NaCN ($K_a$ for HCN $= 6.2 \times 10^{-10}$)
   (e) 0.050 $M$ $HCHO_2$ (formic acid, $K_a = 1.8 \times 10^{-4}$) and 0.025 $M$ $NaCHO_2$ (sodium formate)
   (f) 1.0 $M$ NaOH
   (g) $1.0 \times 10^{-2}$ $M$ $Pu(NO_3)_3$ ($K_a$ for $Pu(OH_2)_6{}^{3+} = 1.1 \times 10^{-7}$)
   (h) 1.00 $M$ HOCl ($K_a = 3.5 \times 10^{-8}$)
   (i) 1.00 $M$ $(CH_3)_3N$ ($K_b = 5.3 \times 10^{-5}$)
   (j) 0.100 $M$ HCl
   (k) $1.0 \times 10^{-3}$ $M$ KOH
   (l) 0.100 $M$ $HNO_2$ ($K_a = 4.0 \times 10^{-4}$)
   0.050 $M$ $NaNO_2$
   (m) 1.00 $M$ HF ($K_a = 7.2 \times 10^{-4}$)
   2.50 $M$ NaF

(n) 0.500 $M$ $CH_3NH_2$ ($K_b = 4.4 \times 10^{-4}$)
(o) $1.00 \times 10^{-2}$ $M$ $Al(NO_3)_3$ ($K_a$ for $Al(OH_2)_6^{3+} = 1.4 \times 10^{-5}$)

2. Calculate the pH for each solution in exercise 1.

3. A solution is to be buffered at pH = 7.0. Which of the following acid-salt pairs would be the best choice for buffering this solution?
   (a) HCl, NaCl
   (b) HOCl ($K_a = 3.5 \times 10^{-8}$), NaOCl
   (c) HCN ($K_a = 6.2 \times 10^{-10}$), NaCN
   (d) HF ($K_a = 7.2 \times 10^{-4}$), NaF

4. To buffer a solution at pH = 7.0, what is the ideal $K_a$ value for the acid to be used?

5. A solution contains 0.50 $M$ propionic acid ($HC_3H_5O_2$, $K_a = 1.3 \times 10^{-3}$) and 0.20 $M$ sodium propionate ($NaC_3H_5O_2$).
   (a) Calculate the pH of this solution.
   (b) Calculate the pH after 0.10 mol of HCl($g$) has been added to 1.0 L of this solution.
   (c) Calculate the pH after 0.050 mol of NaOH($s$) has been added to 1.0 L of the original solution.

6. Calculate the pH of a 0.10 $M$ solution of NaOCN. The $K_a$ for HOCN is $1.3 \times 10^{-4}$.

7. Calculate the pH of a solution formed by mixing 500.0 mL of 0.100 $M$ $NH_3$ and 500.0 mL of 0.0500 $M$ HCl. ($K_b$ for $NH_3$ is $1.8 \times 10^{-5}$.)

8. Consider a solution containing 0.500 $M$ $C_6H_5NH_2$ ($K_b = 3.8 \times 10^{-10}$) and 0.200 $M$ $C_6H_5NH_3Cl$.
   (a) Calculate the pH.
   (b) Calculate the pH after 0.100 mol of HCl($g$) is dissolved in 1.0 L of this solution.

### Verbalizing General Concepts

Answer the following in your own words.

9. What is an acid?

10. What does "acid strength" mean?

11. What is a base?

12. What do strong and weak acids have in common? How are they different?

13. What is a buffered solution?

14. When a salt is dissolved in water, what property of the anion causes the solution to be basic?

15. When a salt is dissolved in water, what properties of the salt might cause the solution to be acidic?

### Multiple Choice Questions

16. The pH of a $10^{-3}$ $M$ NaCl solution is closest to
   (a) 3.0 (b) 6.0
   (c) 7.0 (d) 8.0

17. The acid HX has an ionization constant of $1 \times 10^{-4}$. A solution is 0.1 $M$ in HX and 1 $M$ in the salt $Na^+X^-$. What is a close approximation of the hydrogen ion concentration?
   (a) $10^{-3}$ $M$ (b) $2 \times 10^{-5}$ $M$
   (c) $10^{-5}$ $M$ (d) $10^{-8}$ $M$

**18.** A 0.20 *M* solution of the hypothetical weak acid HZ is found to have a pH of exactly 3.0. The ionization constant, $K_a$, of the acid HZ is
(a) 0.6 (b) $1.0 \times 10^{-3}$
(c) $2.0 \times 10^{-4}$ (d) $5.0 \times 10^{-6}$

**19.** A buffer solution is formed by adding 0.500 mol of sodium acetate and 0.500 mol of acetic acid to 1.00 L $H_2O$. What is the pH of the solution at equilibrium? ($K_a = 1.80 \times 10^{-5}$)
(a) 5.05 (b) 4.74
(c) 4.44 (d) 2.38
(e) None of these

**20.** 0.10 mol HCl is added to the solution in exercise 19. Now what is the concentration of $H^+$?
(a) $2.7 \times 10^{-5}$ *M* (b) $4.5 \times 10^{-5}$ *M*
(c) $3.0 \times 10^{-5}$ *M* (d) $1.2 \times 10^{-5}$ *M*
(e) None of these

**21.** The approximate pH of $10^{-3}$ *M* HCl at 25°C is
(a) $10^{-3}$ (b) $-3$
(c) 3 (d) $10^{3}$
(e) 11

**22.** The $OH^-$ concentration in $10^{-3}$ *M* HCl at 25°C is
(a) $10^{-3}$ *M* (b) $10^{-6}$ *M*
(c) $2 \times 10^{-6}$ *M* (d) $10^{-11}$ *M*
(e) $5 \times 10^{-10}$ *M*

**23.** (a) If $K_b$ for the reaction

$$NH_3(aq) + H_2O(l) \rightleftarrows NH_4^+(aq) + OH^-(aq)$$

is $1.8 \times 10^{-5}$, what concentration of ammonium ion would have to be present at equilibrium in a 0.15 *M* solution of $NH_3$ to make the $OH^-$ concentration $1.4 \times 10^{-4}$ *M*?
(a) 0.15 *M* (b) $1.8 \times 10^{-5}$ *M*
(c) $1.9 \times 10^{-2}$ *M* (d) $1.4 \times 10^{-4}$ *M*
(e) 15 *M*

(b) How would you make the solution in (a) from aqueous ammonia and hydrochloric acid?
(a) 1.0 L of 0.169 *M* $NH_3$ and 1.0 L of 0.019 *M* HCl
(b) 100.0 mL of 0.30 *M* $NH_3$ and 100.0 mL of 0.019 *M* HCl
(c) 1.0 L of 0.15 *M* $NH_3$ and 1.0 L of 0.019 *M* HCl
(d) 500.0 mL of 0.038 *M* HCl and 500.0 mL of 0.338 *M* $NH_3$

**24.** Which species is *most* likely to function *both* as an acid and as a base?
(a) $Cl^-$ (b) $NH_4^+$
(c) $H_2O$ (d) $H_3O^+$

**25.** Under conditions of equal molar concentration in water, which metal ion probably produces the most acidic solution?
(a) $Al^{3+}$ (b) $Ba^{2+}$
(c) $K^+$ (d) $Zn^{2+}$

**26.** Which statement is a logical consequence of the fact that a 0.10 molar solution of potassium acetate, $KC_2H_3O_2$, is less basic than a 0.10 molar solution of potassium cyanide, KCN?
(a) Hydrocyanic acid (HCN) is a weaker acid than acetic acid.
(b) Hydrocyanic acid is less soluble in water than acetic acid.
(c) Cyanides are less soluble than acetates.
(d) Acetic acid is a weaker acid than hydrocyanic acid.

**27.** How many moles of pure NaOH must be used to prepare 10.0 L of a solution that has a pH of 13.00?

(a) 1.0　　(b) 0.10
(c) 0.010　　(d) 0.0010

**28.** What is the pH of a solution that has an $OH^-$ concentration of $4.0 \times 10^{-9}$ *M*?

(a) 8.40　　(b) 5.60
(c) 9.40　　(d) 4.60
(e) None of these

**29.** What is the $H^+$ concentration of a solution produced from 80.0 g of NaOH added to 0.50 L of $H_2O$ (assume no volume increase on addition of NaOH)?

(a) $2.5 \times 10^{-15}$ *M*　　(b) $5.0 \times 10^{-15}$ *M*
(c) $1.0 \times 10^{-14}$ *M*　　(d) $4.0 \times 10^{-14}$ *M*
(e) None of these

**30.** Calculate the $OH^-$ concentration in a 0.0100 *M* solution of aniline, $C_6H_5NH_2$:

$$C_6H_5NH_2(aq) + H_2O(l) \rightleftarrows C_6H_5NH_3^+(aq) + OH^-(aq) \qquad (K_b = 3.8 \times 10^{-10})$$

(a) $3.0 \times 10^{-7}$ *M*　　(b) $3.8 \times 10^{-12}$ *M*
(c) $3.8 \times 10^{-6}$ *M*　　(d) $5.1 \times 10^{-9}$ *M*
(e) None of these

**31.** In this reaction

$$HCN(aq) + HCO_3^-(aq) \rightleftarrows CN^-(aq) + H_2CO_3(aq)$$

$K < 1$ ($K$ is the equilibrium constant). What is the strongest base in this system?

(a) HCN　　(b) $HCO_3^-$
(c) $CN^-$　　(d) $H_2CO_3$

**32.** Consider the reaction

$$CH_3NH_2(aq) + H_2O(l) \rightleftarrows CH_3NH_3^+(aq) + OH^-(aq)$$

where $K_b = 4.4 \times 10^{-4}$. To a solution formed from the addition of 2.0 mol $CH_3NH_2$ to 1.0 L of $H_2O$ is added 1.0 mol of KOH (assume no volume change on addition of solutes). What is the concentration of $CH_3NH_3^+$ at equilibrium?

(a) $3.2 \times 10^{-2}$ *M*　　(b) $2.2 \times 10^{-4}$ *M*
(c) $2.0 \times 10^{-3}$ *M*　　(d) $8.8 \times 10^{-4}$ *M*
(e) None of these

**33.** The Henderson-Hasselbalch equation applied to a solution of an acid, HIn, is

$$pH = pK_a + \log \frac{[In^-]}{[HIn]}$$

You wish to calculate the pH of a solution using the above equation. You know $pK_a$, [HIn], and $[In^-]$. However, you accidentally use twice the concentration of $In^-$ in the calculation. What effect will this have on the calculated pH?

(a) No effect.
(b) Make it too large.
(c) Make it too small.
(d) The pH would be too small or too large depending on the concentration of $In^-$ used in the calculation.

**34.** Which of the following reactions is associated with the normal definition of $K_b$?
(a) $Al(OH_2)_6^{3+}(aq) \rightleftarrows [Al(OH_2)_5OH]^{2+}(aq) + H^+(aq)$
(b) $CN^-(aq) + H^+(aq) \rightleftarrows HCN(aq)$
(c) $F^-(aq) + H_2O(l) \rightleftarrows HF(aq) + OH^-(aq)$
(d) $Cr^{3+}(aq) + 6H_2O(l) \rightleftarrows Cr(OH_2)_6^{3+}(aq)$

**35.** What is the pH of a 1.0 *M* solution of aniline? ($K_b = 4 \times 10^{-10}$)
(a) 4.7 (b) 9.3
(c) 9.4 (d) None of these

**36.** A solution has a pOH of 5.50. What is the $[H^+]$?
(a) 8.5 *M* (b) $10^{-5.5}$ *M*
(c) $10^{-8.5}$ *M* (d) $5.5 \times 10^{-14}$ *M*

**37.** The pH of a 1.0 *M* solution of sodium propionate ($NaC_3H_5O_2$) is 8.3. The $pK_a$ for propionic acid ($HC_3H_5O_2$) is:
(a) 8.3 (b) 5.7
(c) 11.4 (d) 2.6
(e) None of these

**38.** Calculate the $[H^+]$ in a solution prepared by adding 0.050 mol of nitrous acid ($HNO_2$) to 0.100 mol of sodium nitrate ($NaNO_2$) to form 1.00 L of solution. The $K_a$ for $HNO_2$ is $4.0 \times 10^{-4}$.
(a) $2.0 \times 10^{-4}$ *M* (b) $4.0 \times 10^{-4}$ *M*
(c) $8.0 \times 10^{-4}$ *M* (d) $6.3 \times 10^{-3}$ *M*
(e) None of these

**39.** What is the $[H^+]$ after 0.025 mol of HCl(*g*) is added to the solution described in question 38?
(a) $2.0 \times 10^{-4}$ *M* (b) $4.0 \times 10^{-4}$ *M*
(c) $8.0 \times 10^{-4}$ *M* (d) $2.5 \times 10^{-2}$ *M*
(e) None of these

**40.** The $[H^+]$ in a 1.0 *M* solution of HA ($K_a = 4.0 \times 10^{-8}$) is
(a) 1.0 *M* (b) $4.0 \times 10^{-8}$ *M*
(c) $2.0 \times 10^{-8}$ *M* (d) None of these

**41.** What is the pH of a 1.0 *M* solution of HCl?
(a) 10 (b) 1.0
(c) 0 (d) −1.0

**42.** The pH of a 5.0 *M* solution of $HNO_3$ is
(a) 5.0 (b) −0.30
(c) −0.70 (d) 0.70

**43.** If the $pK_a$ of the acid HX is 8.0, the $K_b$ for $X^-$ is
(a) $10^8$ (b) $10^{-8}$
(c) $10^{-6}$ (d) $10^6$
(e) None of these

**44.** The fact that acetic acid ($HC_2H_3O_2$) is a stronger acid than HCN implies that
(a) the acetate ion ($C_2H_3O_2^-$) is a stronger base than $CN^-$.
(b) $CN^-$ is a stronger base than $C_2H_3O_2^-$.
(c) a 0.1 *M* solution of $HC_2H_3O_2$ has a higher pH than 0.1 *M* HCN.
(d) a 0.1 *M* solution of $NaC_2H_3O_2$ has a higher pH than 0.1 *M* NaCN.

**45.** Which of the following is a base-conjugate acid pair?
(a) $C_6H_5NH_2$, $C_6H_5NH_3^+$ (b) $Cl^-$, $NH_4^+$
(c) $NH_3$, $HC_2H_3O_2$ (d) $HC_2H_3O_2$, $H_2O$

**46.** A solution is $1.25 \times 10^{-2}$ $M$ in benzoic acid (HA) and $8.50 \times 10^{-3}$ $M$ in sodium benzoate. If the $K_a$ of benzoic acid is $6.50 \times 10^{-5}$, the $[H^+]$ of this solution is

(a) $1.26 \times 10^{-2}$ $M$ (b) $4.47 \times 10^{-4}$ $M$
(c) $9.56 \times 10^{-5}$ $M$ (d) $4.42 \times 10^{-5}$ $M$
(e) None of these

**47.** A 0.625 $M$ solution of a weak acid (HA) has an $[H^+]$ of $4.20 \times 10^{-3}$. What is the $K_a$ of this weak acid?

(a) $2.82 \times 10^{-5}$ (b) $6.72 \times 10^{-3}$
(c) $1.76 \times 10^{-5}$ (d) $3.55 \times 10^{-4}$
(e) None of these

**48.** When 1.0000 g of sodium acetate (MW = 82.0) is added to 150.0 mL of 0.0500 molar acetic acid (MW = 60.0), the pOH of the resulting solution is ($K_a$ for acetic acid is $1.8 \times 10^{-5}$)

(a) 1.06 (b) 4.97
(c) 9.04 (d) 12.90
(e) None of these

**49.** The amount (in grams) of sodium acetate (MW = 82.0) to be added to 500.0 mL of 0.200 molar acetic acid ($K_a = 1.80 \times 10^{-5}$) in order to make a buffer with pH = 5.000 is

(a) 369 (b) 0.180
(c) 14.8 (d) 29.5
(e) None of these

**50.** A solution is 0.120 $M$ in acetic acid and 0.0900 $M$ in sodium acetate. Calculate the $[H^+]$ at equilibrium. ($K_a$ of acetic acid is $1.80 \times 10^{-5}$)

(a) 4.87 $M$ (b) $2.40 \times 10^{-5}$ $M$
(c) $1.35 \times 10^{-5}$ $M$ (d) 4.62 $M$
(e) $6.00 \times 10^{-6}$ $M$

**51.** A 0.30 $M$ solution of a weak acid (HA) has an $[H^+]$ of $1.66 \times 10^{-4}$ $M$. What is the $K_a$ of this weak acid?

(a) $4.8 \times 10^{1}$ (b) $5.5 \times 10^{-4}$
(c) $1.2 \times 10^{8}$ (d) $9.2 \times 10^{-8}$
(e) The $K_a$ cannot be calculated without additional information

**52.** What volumes of 0.500 $M$ $HNO_2$ ($K_a = 4.0 \times 10^{-4}$) and 0.500 $M$ $NaNO_2$ must be mixed to prepare 1.0 L of a solution buffered at pH 3.55?

(a) 500 mL of each solution
(b) 703 mL 0.500 $M$ $HNO_2$, 297 mL 0.500 $M$ $NaNO_2$
(c) 413 mL 0.500 $M$ $HNO_2$, 587 mL 0.500 $M$ $NaNO_2$
(d) 297 mL 0.500 $M$ $HNO_2$, 703 mL 0.500 $M$ $NaNO_2$
(e) 587 mL 0.500 $M$ $HNO_2$, 413 mL 0.500 $M$ $NaNO_2$

# 3 Acid-Base Titrations and Indicators

## CHAPTER OBJECTIVES

1. Define titration, millimole, indicator, equivalence (stoichiometric) point, endpoint.
2. Calculate the pH of a solution during the titration of a strong acid with a sodium hydroxide solution.
3. Calculate the pH of a solution during the titration of a weak acid with a sodium hydroxide solution.
4. Know the differences between the titration curves for the titration of a strong acid with a strong base and that of a weak acid with a strong base.
5. Calculate pH when strong or weak bases are titrated with solutions of strong acids.
6. Know the mechanism for indicator color changes.
7. Know the principles involved in choosing an appropriate indicator.

## 3.1 Introduction

An acid-base titration involves a quantitative study of the reaction occurring when a solution containing a base is mixed with a solution containing an acid. The calculations involved in a titration depend on concepts already covered, but their application may not be completely straightforward. Thus it is especially important to attack these problems in a systematic way.

A titration problem always involves a stoichiometry problem, and may also involve an equilibrium problem. The two problems must be treated separately.

The simplest case, the titration involving a strong acid and a strong base, will be treated first.

## 3.2 Strong Acid–Strong Base Titrations

This case involves only a stoichiometry problem. No equilibrium calculations are necessary.

**Sample Exercise 3.1** Consider the titration of 100.0 mL of 1.00 *M* HCl with 0.500 *M* NaOH. Calculate the $[H^+]$ in the solution after 50.0 mL of 0.50 *M* NaOH has been added.

**Solution**

A titration always involves a chemical reaction. To determine what reaction will occur, consider the major species existing in the mixed solution before any reaction occurs:

$$H^+, Cl^-, Na^+, OH^-, H_2O$$

Note that the solution contains large quantities of $H^+$ and $OH^-$. These ions readily react to form $H_2O$:

$$H^+(aq) + OH^-(aq) \rightleftarrows H_2O(l)$$

Since this equilibrium lies far to the right ($K = 1/K_w = 10^{14}$), assume the reaction that occurs upon mixing goes to completion (that is, the reaction proceeds until it "runs out" of the limiting reagent). Now do a stoichiometry problem:

How much $H^+$ was originally present in the solution? 100.0 mL of 1.00 *M* HCl gives

$$100.0 \text{ mL} \times \frac{1.0 \text{ L}}{1000 \text{ mL}} \times 1.00 \frac{\text{mol}}{\text{L}} = 0.100 \text{ mol HCl}$$

$$= 0.100 \text{ mol } H^+$$

since the HCl (a strong acid) is completely dissociated.

How much NaOH was added? 50.0 mL of 0.50 *M* NaOH was added, and

$$50.0 \text{ mL} \times \frac{1.0 \text{ L}}{1000 \text{ mL}} \times 0.50 \frac{\text{mol}}{\text{L}} = 0.025 \text{ mol NaOH}$$

$$= 0.025 \text{ mol } OH^-$$

since the NaOH (a strong base) is completely dissociated.

The reaction is

| | $H^+$ | + | $OH^+$ | → | $H_2O$ |
|---|---|---|---|---|---|
| Before the reaction | 0.100 mol | | 0.025 mol | | large and essentially constant |
| After the reaction goes to completion | 0.100 − 0.025 = 0.075 mol | | 0.025 − 0.025 = 0 | | large and essentially constant |

After the reaction is complete, the major species present in the solution are

$$H^+, Cl^-, Na^+, H_2O$$

The $[H^+]$ will be determined by the excess $H^+$ remaining:

$$[H^+] = \frac{\text{mol}}{\text{volume(L)}} = \frac{0.075 \text{ mol}}{0.100 \text{ L} + 0.050 \text{ L}} = \frac{0.075 \text{ mol}}{0.150 \text{ L}}$$

↗ volume of HCl solution ↗ volume of NaOH solution ↗ total volume of the mixture

$$[H^+] = 0.50 \, M$$

$$\text{pH} = -\log(0.50) = 0.30$$

Titration problems typically involve volumes of 100 mL or less. Having to change the units on these volumes to liters is a hassle and a source of "silly errors." This problem can be avoided by using the units milliliters and millimoles. A millimole (mmol) is one one-thousandth of a mole (mol/1000, or $10^{-3}$ mol):

$$\text{Molarity} = \frac{\text{mol}}{\text{L}} = \frac{\frac{\text{mol}}{1000}}{\frac{\text{L}}{1000}} = \frac{\text{mmol}}{\text{mL}}$$

Thus molarity can be expressed as either mol/L or mmol/mL. The volume in *m*L times the molarity gives the number of millimoles. For example,

100.0 mL of 1.00 $M$ HCl contains

$$100.0 \text{ mL} \times 1.00 \frac{\text{mmol HCl}}{\text{mL}} = 100 \text{ mmol HCl}$$

$$50.0 \text{ mL} \times \frac{0.50 \text{ mmol}}{\text{mL}} = 25 \text{ mmol NaOH}$$

Now do the calculations for sample exercise 3.1 using millimoles:

| | $H^+$ | + | $OH^-$ | $\longrightarrow$ | $H_2O$ |
|---|---|---|---|---|---|
| Before the reaction | 100 mmol | | 25 mmol | | |
| After the reaction goes to completion | 75 mmol | | 0 | | |

$$[H^+] = \frac{\text{mmol } H^+}{\text{mL of solution}} = \frac{75 \text{ mmol}}{(100 + 50) \text{ mL}} = \frac{75 \text{ mmol}}{150 \text{ mL}}$$

$$= 0.50 \frac{\text{mmol}}{\text{mL}}$$

$$\frac{0.50 \text{ mmol}}{\text{mL}} = \frac{0.50 \text{ mol}}{\text{L}} = 0.50\ M$$

The advantage in using millimoles to do titration problems is that it eliminates the need to change the volume to liters, which can lead to math errors. Remember:

$$\text{mmol} = \text{mL} \times \text{molarity}$$

$$\text{mol} = \text{L} \times \text{molarity}$$

Learn to use millimoles when the volumes are given in milliliters. Use moles when the volumes are given in liters.

**Sample Exercise 3.2** Consider the titration of 50.0 mL of 0.200 $M$ $HNO_3$ with 0.100 $M$ NaOH. Calculate the pH at the following points in the titration:

*Total volume of 0.100 M NaOH added (mL)*

**A.** 0
**B.** 10.0
**C.** 20.0

**D.** 50.0
**E.** 100.0
**F.** 150.0
**G.** 200.0

**Solution**

**A.** 0 mL of NaOH added.

The solution contains $HNO_3$, $H_2O$. $HNO_3$ is a strong acid and thus is completely dissociated:

$$HNO_3(aq) \longrightarrow H^+(aq) + NO_3^-(aq)$$

0.20 $M$ $HNO_3$ produces 0.20 $M$ $H^+$, so $[H^+] = 0.20\ M$, and

$$pH = -\log(0.20) = 0.70$$

**B.** 10.0 mL of NaOH added.

Before the reaction, the mixed solution contains

$$HNO_3,\ NaOH,\ H_2O$$

Since both the $HNO_3$ and NaOH are completely dissociated, the solution would contain

$$H^+,\ NO_3^-,\ Na^+,\ OH^-,\ H_2O$$

Note that large quantities of both $H^+$ and $OH^-$ are present. These ions will immediately react to form water:

| | $H^+$ | + | $OH^-$ | $\longrightarrow$ | $H_2O$ |
|---|---|---|---|---|---|
| Before the reaction | 50.0 mL × 0.200 $M$ = 10.0 mmol | | 10.0 mL × 0.10 $M$ = 1.0 mmol | | large and constant |
| After the reaction | 10.0 − 1.0 = 9.0 mmol | | 1.0 − 1.0 = 0 | | large and constant |

After the reaction the solution contains

$$H^+,\ NO_3^-,\ Na^+,\ H_2O \text{ (the } OH^- \text{ has been consumed)}$$

$$[H^+] = \frac{\text{no. of mmol } H^+}{\text{volume of solution (mL)}} = \frac{9.0}{50.0 + 10.0} = 0.15\ M$$

↗ volume of acid solution ↖ volume of NaOH added

$$pH = -\log(0.15) = 0.82$$

**C.** 20.0 mL (total) of NaOH added.

Consider this problem as 20.0 mL of NaOH added to the original solution rather than as 10.0 mL of NaOH added to the solution we have just dealt with in B. The problem can be done either way, but it is best to go back to the original solution each time, so that a mistake made in an earlier part does not invalidate each succeeding calculation. As before, the added $OH^-$ will react with $H^+$:

| | $H^+$ | + | $OH^-$ | $\longrightarrow$ | $H_2O$ |
|---|---|---|---|---|---|
| Before the reaction | 50.0 mL × 0.200 $M$ = 10.0 mmol | | 20.0 mL × 0.10 $M$ = 2.0 mmol | | |
| After the reaction | 10.0 − 2.0 = 8.0 mmol | | 2.0 − 2.0 = 0 mmol | | |

After the reaction,

$$[H^+] = \frac{8.0 \text{ mmol } H^+}{50.0 + 20.0 \text{ mL}} = 0.11\ M$$

$$pH = -\log(0.11) = 0.94$$

**D.** 50.0 mL (total) of NaOH added.

Do this one to test what you have learned. (Answer: pH = 1.30.)

**E.** 100.0 mL (total) of NaOH added.

At this point the amount of NaOH added is

$$100.0 \text{ mL} \times 0.100\ M = 10.0 \text{ mmol}$$

The original amount of nitric acid was

$$50.0 \text{ mL} \times 0.200\ M = 10.0 \text{ mmol}$$

Enough $OH^-$ has been added to react exactly with the $H^+$ from the nitric acid. This is the **stoichiometric point** or the **equivalence point** of the titration. At this point, the main species in solution are $Na^+$, $NO_3^-$, $H_2O$. Since neither $Na^+$ nor $NO_3^-$ is an acid or a base ($NO_3^-$ is the anion of the strong acid $HNO_3$ and thus is a *very* weak base), the solution is neutral; the pH is 7.0.

**F.** 150.0 mL (total) of NaOH added.

Now the $OH^-$ added is in excess. The reaction is

| | $H^+$ | + | $OH^-$ | $\longrightarrow$ $H_2O$ |
|---|---|---|---|---|
| Before the reaction | $50.0 \text{ mL} \times 0.200\ M$ = 10.0 mmol | | $150.0 \text{ mL} \times 0.100\ M$ = 15.0 mmol | |
| After the reaction | $10.0 - 10.0$ = 0 | | $15.0 - 10.0$ = 5.0 mmol | |

The $OH^-$ is in excess, and it will determine the pH.

$$[OH^-] = \frac{\text{no. of mmol } OH^-}{\text{volume (mL)}} = \frac{5.0 \text{ mmol}}{50.0 + 150.0 \text{ mL}} = \frac{5.0 \text{ mmol}}{200.0 \text{ mL}} = 0.025\ M$$

The $[H^+]$ can be found by substituting into the water equilibrium:

$$[H^+][OH^-] = [H^+](0.025\ M) = 1.0 \times 10^{-14}$$

$$[H^+] = \frac{1.0 \times 10^{-14}}{2.5 \times 10^{-2}} = 4.0 \times 10^{-13}\ M$$

$$pH = -\log(4.0 \times 10^{-13}) = 12.40$$

**G.** 200.0 mL of NaOH added to the original solution.

Do this one to test your understanding. (Answer: pH = 12.60.)

A graph of a titration where pH is plotted against the volume of titrant added is called a **pH curve**. The pH curve for the titration considered above is shown in Figure 3.1.This pH curve is typical of the titration of a strong acid with a sodium hydroxide solution.

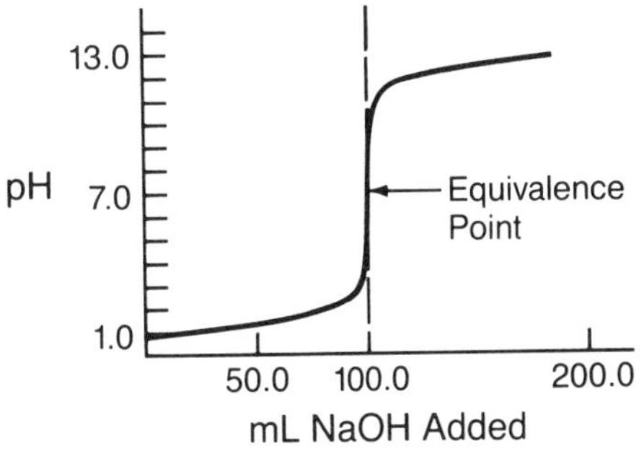


**Figure 3.1**

## Summary

The following are characteristics of the calculations involved in the titration of a strong acid with a strong base ($OH^-$):

1. Before the stoichiometric point, calculate the number of millimoles of acid left and divide by the total volume of the solution to obtain the $[H^+]$.
2. At the stoichiometric (equivalence) point, the pH = 7.0.
3. After the stoichiometric point, calculate the number of millimoles of $OH^-$ in excess and divide by the total volume of the solution to obtain the $[OH^-]$. Then use $K_w$ to obtain the $[H^+]$.

The titration of a strong base with a strong acid is the opposite of the case just considered.

---

**TEST 3.1** Consider the titration of 100.0 mL of 0.50 *M* NaOH with 1.0 *M* HCl. Calculate the pH at the following total volumes of added HCl:

**A.** 0.0 mL
**B.** 25.0 mL
**C.** 50.0 mL
**D.** 75.0 mL

---

# 3.3 Weak Acid–Strong Base Titrations

Titrations considered so far have involved strong acids and strong bases. Now titrations involving weak acids and strong bases will be discussed. In these cases, two problems will have to be solved:

1. A stoichiometry problem.
2. An equilibrium problem.

IT IS ESSENTIAL TO DO THESE PROBLEMS SEPARATELY.

As an illustration, consider the following situation: 30.0 mL of 0.10 *M* NaOH solution is added to 50.0 mL of 0.10 *M* HF. Calculate the pH in the mixed solution. ($K_a$ for HF is $7.2 \times 10^{-4}$.)

Consider the major components of the mixed solution *before any reaction occurs:*

$$HF, OH^-, Na^+, H_2O$$

What reaction will occur? Note that the solution contains large quantities of a weak acid, HF, and a very strong base, $OH^-$. Will $OH^-$ be able to take $H^+$ from HF? That is, will the reaction

$$OH^-(aq) + HF(aq) \longrightarrow H_2O(l) + F^-(aq)$$

occur?

The answer is a definite "yes." $OH^-$ is a much stronger base than $F^-$. The equilibrium constant for the above reaction is $7.2 \times 10^{10}$, which is a very large number. For all practical purposes one may make the assumption that the reaction between HF and $OH^-$ goes to completion (until one of them runs out). This is typical of weak acids: *$OH^-$ can be assumed to react completely with all common weak acids.*

The first step in a titration problem is to do the stoichiometry calculations. In this case, the following questions must be answered:

1. What major species are present in solution before any reaction occurs?
2. What reaction occurs?
3. How much HF was originally present?
4. How much $OH^-$ was added?
5. How much HF is consumed?
6. How much $F^-$ is formed?

Consider these questions one at a time:

1. The solution (before any reaction) contains HF, $Na^+$, $OH^-$, $H_2O$.
2. The $OH^-$ will react to completion with the HF:

$$OH^- + HF \longrightarrow H_2O + F^-$$

3. The amount of HF originally present is

$$50.0 \text{ mL} \times 0.10\ M = 5.0 \text{ mmol}$$

4. The amount of $OH^-$ added is

$$30.0 \text{ mL} \times 0.10\ M = 3.0 \text{ mmol}$$

5. The $OH^-$ will consume 3.0 mmol of HF.
6. 3.0 mmol of $F^-$ will be formed.

All of this can be summarized as follows:

| | HF | + | $OH^-$ | $\longrightarrow$ | $F^-$ | + | $H_2O$ |
|---|---|---|---|---|---|---|---|
| Before the reaction | 5.0 mmol | | 3.0 mmol | | 0 | | |
| After the reaction | 5.0 − 3.0 = 2.0 mmol | | 3.0 − 3.0 = 0 | | 3.0 mmol formed | | |

After the reaction goes to completion, the solution contains the major species

$$HF, F^-, Na^+, H_2O$$

To calculate the pH, an equilibrium problem must now be done. Which equilibrium makes the dominant contribution to the $[H^+]$? Looking at the above collection of species, it is clear that the HF dissociation (which involves both HF and $F^-$) will control the pH. Do the equilibrium problem in the usual way:

$$HF(aq) \rightleftarrows H^+(aq) + F^-(aq) \qquad K_a = \frac{[H^+][F^-]}{[HF]} = 7.2 \times 10^{-4}$$

The volume of the solution is 50.0 + 30.0 = 80.0 mL. Remember that the *total volume* of the mixed solutions must be used in carrying out the equilibrium calculations.

*Initial Concentrations (before any dissociation)*

$$[HF]_0 = \frac{2.0 \text{ mmol HF left}}{(50.0 + 30.0) \text{ mL}}$$

$$[F^-]_0 = \frac{3.0 \text{ mmol } F^- \text{ formed}}{(50.0 + 30.0) \text{ mL}}$$

(Formed in the reaction of $OH^-$ with HF)

$$[H^+]_0 \approx 0$$

Let $x$ mol/L of HF dissociate →

*Equilibrium Concentrations*

$$[HF] = \frac{2.0}{80.0} - x$$

$$[F^-] = \frac{3.0}{80.0} + x$$

$$[H^+] = x$$

Note that this is really a "common ion problem." The common ion is the $F^-$ formed in the initial, complete reaction of $OH^-$ with HF.

Now substitute into the equilibrium expression and solve for $x$:

$$7.2 \times 10^{-4} = K_a = \frac{[H^+][F^-]}{[HF]} = \frac{(x)\left(\frac{3.0}{80.0} + x\right)}{\left(\frac{2.0}{80.0} - x\right)} \approx \frac{(x)\left(\frac{3.0}{80.0}\right)}{\left(\frac{2.0}{80.0}\right)}$$

$$x \approx \frac{(7.2 \times 10^{-4})\left(\frac{2.0}{80.0}\right)}{\left(\frac{3.0}{80.0}\right)} = 4.8 \times 10^{-4} \, M$$

Since $2.0/80.0 = 2.5 \times 10^{-2}$ and $x = 4.8 \times 10^{-4}$, the assumptions that $2.0/80.0 - x \approx 2.0/80.0$ and $3.0/80.0 + x \approx 3.0/80.0$ are valid.

$$x = [H^+] = 4.8 \times 10^{-4}$$

$$pH = -\log[H^+] = -\log(4.8 \times 10^{-4}) = 3.32$$

For the titration of a weak acid with a strong base, the following steps are necessary:

1. Calculate the amount (in millimoles) of weak acid (HA) originally in solution.
2. Calculate the number of millimoles of $OH^-$ added.
3. Run the reaction between $OH^-$ and the weak acid to completion:

$$HA + OH^- \longrightarrow H_2O + A^-$$

4. Calculate the amounts (millimoles) of HA left and of $A^-$ formed.
5. Do the weak acid equilibrium problem as usual. (Do not forget to account for the $A^-$ formed in the reaction with $OH^-$ when computing the initial concentrations.)

---

**Sample Exercise 3.3**

Consider the titration of 50.0 mL of 0.10 $M$ acetic acid (HOAc) with 0.10 $M$ NaOH. ($K_a$ for HOAc is $1.8 \times 10^{-5}$.) Calculate the pH at each of the following total added volumes of NaOH.

**Solution**

**A.** 0 mL.

This is a standard weak acid calculation. Do it for practice. (Answer: pH = 2.87.)

**B.** 10.0 mL of 0.10 $M$ NaOH added.

Consider the major species in the mixed solution before any reaction takes place. (*Always* do this. The hardest things about acid-base problems are to decide what reaction takes place and what equilibrium dominates. Writing the major species is the key to making the correct decisions. The importance of doing this cannot be overemphasized.)

In solution, the major species are HOAc, $OH^-$, $Na^+$, $H_2O$. The strong base, $OH^-$, will react with the strongest proton donor, which in this solution is HOAc. As always, do the stoichiometry problem first.

| | $OH^-$ | + | HOAc | $\longrightarrow$ | $OAc^-$ | + $H_2O$ |
|---|---|---|---|---|---|---|
| Before the reaction | 10.0 mL × 0.10 *M* = 1.0 mmol | | 50.0 mL × 0.10 *M* = 5.0 mmol | | 0 | |
| After the reaction | 0 | | 5.0 − 1.0 = 4.0 mmol | | 1.0 mmol | |

Now, to decide what equilibrium to use, examine the major components left in the solution after the reaction:

$$HOAc, OAc^-, Na^+, H_2O$$

1. HOAc Weak acid
2. $OAc^-$ Conjugate base of HOAc
3. $Na^+$ No effect on pH
4. $H_2O$ Very weak acid/base

The pH will be determined by the acetic acid equilibrium:

$$HOAc(aq) \rightleftarrows H^+(aq) + OAc^-(aq)$$

which involves HOAc and $OAc^-$ and thus reflects the effects of each of these components on the pH of the solution. Now follow the usual steps to complete the equilibrium calculations:

*Initial Concentrations*

$$[HOAc]_0 = \frac{4.0 \text{ mmol}}{(50.0 + 10.0) \text{ mL}} = \frac{4.0}{60.0}$$

$$[OAc^-]_0 = \frac{1.0 \text{ mmol}}{(50.0 + 10.0) \text{ mL}} = \frac{1.0}{60.0}$$

$$[H^+]_0 \approx 0$$

*Let x mol/L of HOAc dissociate* →

*Equilibrium Concentrations*

$$[HOAc] = \frac{4.0}{60.0} - x$$

$$[OAc^-] = \frac{1.0}{60.0} + x$$

$$[H^+] = x$$

$$1.8 \times 10^{-5} = K_a = \frac{[H^+][OAc^-]}{[HOAc]} = \frac{(x)\left(\frac{1.0}{60.0} + x\right)}{\left(\frac{4.0}{60.0} - x\right)} \approx \frac{x\left(\frac{1.0}{\cancel{60.0}}\right)}{\left(\frac{4.0}{\cancel{60.0}}\right)}$$

$$x = \left(\frac{4.0}{1.0}\right)(1.8 \times 10^{-5}) = 7.2 \times 10^{-5} = [H^+]$$

$$pH = -\log(7.2 \times 10^{-5}) = 4.14$$

The next step in the titration will be done in the format of a test.

**TEST 3.2** C. 25.0 mL of 0.10 *M* NaOH has been added to the original solution.

1. List the major species before any reaction occurs.
2. Decide what reaction occurs.

3. Do the stoichiometry problem.
   (a) How much HOAc remains after the reaction is complete?
   (b) How much $OAc^-$ is formed by the reaction?
4. Do the equilibrium problem.
   (a) List the major species after the reaction of question 2 occurs.
   (b) Decide which equilibrium will dominate.
   (c) Write down the initial concentrations (after the reaction of question 2 goes to completion but before any HOAc dissociation occurs).
   (d) Define $x$.
   (e) Write the equilibrium concentrations in terms of $x$.
   (f) Solve for $x$.
   (g) Calculate the pH.

---

**Sample Exercise 3.3 continued**

This point in the titration is a special one. It is halfway to the equivalence point. The original solution consisted of 50.0 mL of 0.10 *M* HOAc, which contains 5.0 mmol of HOAc.

What amount of $OH^-$ is required to reach the equivalence point? Remember that the equivalence point occurs when the amount of $OH^-$ added is equal to the total amount of acid. (Thus, in this problem, 5.0 mmol of $OH^-$ is required.) Since

$$50.0 \text{ mL} \times 0.100\ M = 5.0 \text{ mmol of } OH^-$$

50.0 mL of NaOH is required to reach the stoichiometric point. Thus 25.0 mL of added base represents the halfway point (half of the original HA has been converted to $A^-$) where $[HA] \approx [A^-]$. This leads to the conclusion that $[H^+] = K_a$, which can be shown as follows:

$$1.8 \times 10^{-5} \frac{[H^+][OAc^-]}{[HOAc]} = \frac{(x)\left(\frac{2.5}{75.0} + x\right)}{\frac{2.5}{75.0} - x} \approx \frac{(x)\frac{2.5}{75.0}}{\frac{2.5}{75.0}}$$

$$x = 1.8 \times 10^{-5} = K_a$$

This is an important result. *At the point halfway to the equivalence point,*

$$[H^+] = K_a$$
$$-\log [H^+] = -\log K_a$$
$$pH = pK_a$$

**D.** 40.0 mL (total) of NaOH added.

Do this one to test yourself. (Answer: pH = 5.35.)

**E.** 50.0 mL (total) of NaOH added.

This is the equivalence point of the titration, 5.0 mmol of $OH^-$ has been added, which will just react with the 5.0 mmol of HOAc originally present. At this point (*after the* $OH^-$, HOAc *reaction is complete*) the solution contains

$$NA^+, OAc^-, H_2O$$

How is the pH calculated? Note that the solution contains $OAc^-$, which is a base. Remember that a base will combine with a proton. The only source of protons in the solution is $H_2O$, so that the reaction that will occur is

$$OAc^-(aq) + H_2O(l) \rightleftarrows HOAc(aq) + OH^-(aq)$$

$$K_b = \frac{[HOAc][OH^-]}{[OAc^-]} = \frac{K_w}{K_a} = \frac{1.0 \times 10^{-14}}{1.8 \times 10^{-5}} = 5.6 \times 10^{-10}$$

*Initial Concentrations (before any $OAc^-$ reacts)*

$$[OAc^-]_0 = \frac{5.0 \text{ mmol}}{(50.0 + 50.0) \text{ mL}}$$

$$[OH^-]_0 \approx 0$$

$$[HOAc]_0 = 0$$

$$\xrightarrow[\text{with } H_2O]{\text{Let } x \text{ mol/L } OAc^- \text{ react}}$$

*Equilibrium Concentrations*

$$[OAc^-] = \frac{5.0}{100} - x$$

$$[OH^-] = x$$

$$[HOAc] = x$$

$$5.6 \times 10^{-10} = K_b = \frac{[HOAc][OH^-]}{[OAc^-]} = \frac{(x)(x)}{0.050 - x} \approx \frac{x^2}{0.050}$$

$$x^2 \approx (0.050)(5.6 \times 10^{-10}) = 2.8 \times 10^{-11}$$

$$x \approx 5.3 \times 10^{-6}$$

The assumption, $0.050 - x = 0.050$, is valid, so

$$[OH^-] = 5.3 \times 0^{-6} \ M$$

$$1.0 \times 10^{-14} = K_w \ [H^+][OH^-] = [H^+](5.3 \times 10^{-6} \ M)$$

$$[H^+] = 1.9 \times 10^{-9} \ M$$

$$pH = -\log(1.9 \times 10^{-9}) = 8.72$$

This is another important result: The pH at the stoichiometric point in the titration of a weak acid with a strong base is always greater than 7. This results because the anion of the acid, which remains in solution at the stoichiometric point, is a base.

Contrast this weak acid–strong base titration to the titration of a strong acid with a strong base, where the pH at the stoichiometric point is 7.0 because the anion remaining is not a good base.

**F.** 60.0 mL (total) of NaOH added.

At this point, excess $OH^-$ has been added:

| | $OH^-$ | + | $HOAc$ | $\longrightarrow$ | $OAc^-$ | + | $H_2O$ |
|---|---|---|---|---|---|---|---|
| Before the reaction | $60.0 \times 0.10\ M$ = 6.0 mmol | | $50.0 \text{ mL} \times 0.10\ M$ = 5.0 mmol | | 0 | | |
| After the reaction | 6.0 − 5.0 = 1.0 mmol | | 5.0 − 5.0 = 0 | | 5.0 mmol | | |

After the reaction, the solution contains $Na^+$, $OAc^-$, $OH^-$, and $H_2O$. The pH will be determined by the excess $OH^-$:

$$[OH^-] = \frac{\text{no. of mmol } OH^-}{\text{volume (mL)}} = \frac{1.0 \text{ mmol}}{(50.0 + 60.0) \text{ mL}}$$

$$= 9.1 \times 10^{-3} \ M$$

$$1.0 \times 10^{-14} = [H^+][OH^-] = [H^+](9.1 \times 10^{-3})$$

$$[H^+] = \frac{1.0 \times 10^{-14}}{9.1 \times 10^{-3}} = 1.1 \times 10^{-12} \ M$$

$$pH = 11.96$$

**G.** 75.0 mL of NaOH added.

Do this one to test yourself. (Answer: pH = 12.30.)

The pH curve for the titration considered above is shown in Figure 3.2. This curve is typical of the titration of a weak acid with a strong base. Note the difference in shape between this curve and that for the titration of a strong acid (Figure 3.1). Near the beginning of the titration of a weak acid, the pH changes more rapidly than in the middle region of titration because of the buffering effect, which is greatest when $[HA] = [A^-]$ at the halfway point (25 mL of NaOH). Note that the rate of change of pH is smallest in this region.

The other notable difference between the curves for strong and weak acids is the pH corresponding to the equivalence point. For the titration of a strong acid, the equivalence point occurs at pH 7. For a weak acid, on the other hand, the equivalence point occurs at a pH greater than 7 because of the basicity of the conjugate base of the weak acid.

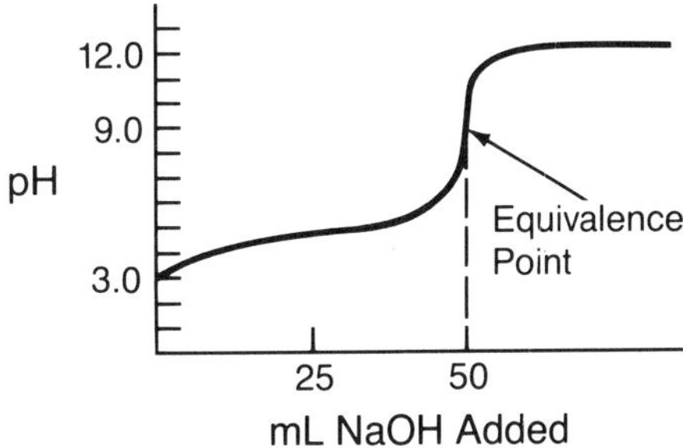


**Figure 3.2**

**Sample Exercise 3.4**

A 0.350 g sample of a solid weak acid (HA) is dissolved in 50.0 mL of water and titrated with 0.100 *M* NaOH.

**A.** 23.2 mL of 0.100 *M* NaOH is required to reach the equivalence point. What is the molecular weight of the acid?

**Solution**

The titration reaction is

$$HA(aq) + OH^-(aq) \longrightarrow A^-(aq) + H_2O(l)$$

23.2 mL × 0.100 *M* NaOH = 2.32 mmol of $OH^-$ have been added. This means that 2.32 mmol of HA must have been dissolved in the solution originally. Thus

$$0.350 \text{ g of HA} = 2.32 \text{ mmol HA} = 2.32 \times 10^{-3} \text{ mol HA}$$

$$\text{MW of HA} = \frac{0.350 \text{ g}}{2.32 \times 10^{-3} \text{ mol}} = 151 \text{ g/mol}$$

**B.** After 10.0 mL of 0.100 *M* NaOH was added, the pH was measured and found to be 4.00. Calculate the $K_a$ for the weak acid.

**Solution**

From part A, the solution originally contained 2.32 mmol of HA. 10.0 mL of 0.100 *M* NaOH contains 10.0 mL × 0.10 *M* = 1.00 mmol of $OH^-$. The reaction is

| | HA | + | $OH^-$ | ⟶ | $A^-$ | + $H_2O$ |
|---|---|---|---|---|---|---|
| Before the reaction | 2.32 mmol | | 1.00 mmol | | 0 | |
| After the reaction | 2.32 − 1.00 = 1.32 mmol | | 1.00 − 1.00 = 0 mmol | | 1.00 mmol | |

Now consider the dissociation of the HA that remains:

$$HA(aq) \rightleftarrows H^+(aq) + A^-(aq) \qquad K_a = \frac{[H^+][A^-]}{[HA]}$$

*Initial Concentrations (before dissociation)*

$$[HA]_0 = \frac{1.32 \text{ mmol}}{(50.0 + 10.0) \text{ mL}}$$

$$[A^-]_0 = \frac{1.00 \text{ mmol}}{(50.0 + 10.0) \text{ mL}}$$

$$[H^+]_0 \approx 0$$

Let $x$ mol/L HA dissociate →

*Equilibrium Concentrations*

$$[HA] = \frac{1.32}{60.0} - x$$

$$[A^-] = \frac{1.00}{60.0} + x$$

$$[H^+] = x$$

In this problem, we know that the pH = 4.00. Since $[H^+] = 10^{-pH}$, the $[H^+]$ is $1.0 \times 10^{-4}$ *M*.

$$K_a = \frac{[H^+][A^-]}{[HA]} = \frac{(1.0 \times 10^{-4})\left(\frac{1.00}{60.0} + 1.0 \times 10^{-4}\right)}{\frac{1.32}{60.0} - 1.0 \times 10^{-4}}$$

$$= \frac{(1.0 \times 10^{-4})(1.67 \times 10^{-2} + 1.0 \times 10^{-4})}{2.20 \times 10^{-2} - 1.0 \times 10^{-4}}$$

$$K_a \approx \frac{(1.0 \times 10^{-4})(1.67 \times 10^{-2})}{2.20 \times 10^{-2}} = 7.6 \times 10^{-5}$$

**TEST 3.3**

**A.** Consider 25.0 mL of 0.30 *M* $HNO_2$ ($K_a = 4.0 \times 10^{-4}$) that is titrated with 0.500 *M* NaOH.

1. What volume of NaOH solution is required to reach the equivalence point?
2. Calculate the pH of the solution after the following total volumes of base are added.
   (a) 5.0 mL (b) 7.5 mL
   (c) 15.0 mL (d) 20.0 mL

**B.** 0.200 g of an unknown solid acid (HA) is dissolved in 100.0 mL of water.

1. 40.0 mL of 0.0500 *M* NaOH was required to reach the stoichiometry point. What is the molecular weight of the acid?
2. The pH of the above solution after 20.0 mL of 0.0500 *M* NaOH has been added in 6.00. What is the $K_a$ for this acid?
3. Calculate the pH of the solution at the equivalence point for the titration of the original solution containing HA with 0.0500 *M* NaOH.

## 3.4 Titration of Weak Bases with Strong Acids

Since all of the chemistry needed to handle this case has already been introduced, this section will encourage you to generalize your knowledge by doing a problem in the format of a test.

**TEST 3.4** Consider the titration of 100.0 mL of 0.050 $M$ $NH_3$ with 0.10 $M$ HCl. The $K_b$ for $NH_3$ is $1.8 \times 10^{-5}$.

Calculate the pH at the following total volumes of added HCl:

**A.** 0 mL added.

**B.** 10.0 mL has been added to the original solution. Before any reaction occurs, the solution contains $NH_3$, $H^+$, $Cl^-$, and $H_2O$. The reaction will be

$$NH_3(aq) + H^+(aq) \longrightarrow NH_4^+(aq)$$

which can be assumed to go to completion.

1. Do the stoichiometry problem.
   (a) How much $NH_3$ is consumed in the reaction?
   (b) How much $NH_3$ remains after the reaction?
   (c) How much $NH_4^+$ is formed in the reaction?
2. Do the equilibrium problem.
   (a) What major species are present?
   (b) What equilibrium will be dominant?
   (c) List the initial concentrations.
   (d) Define $x$.
   (e) Write the equilibrium concentrations in terms of $x$.
   (f) Solve for $x$.
   (g) Calculate the pH.

**C.** 25.0 mL of HCl has been added to the original solution.

**D.** 50.0 mL of HCl has been added to the original solution. This is the stoichiometric point. After the titration is complete, the solution contains $NH_4^+$, $Cl^-$, and $H_2O$.

1. What equilibrium will be dominant?
2. Calculate the pH.

**E.** 60.0 mL of HCl has been added to the original solution. The solution now contains excess $H^+$. Calculate the $[H^+]$ and the pH.

## 3.5 Indicators

An acid-base indicator is a substance that indicates the endpoint of an acid-base titration by changing color. The most commonly used acid-base indicators are complex organic molecules that are weak acids, HIn. They exhibit one color when a proton is attached and a different color when the proton is absent. To see how these molecules function as indicators, consider the following weak acid equilibrium:

$$\underset{\text{red}}{HIn(aq)} \rightleftarrows H^+(aq) + \underset{\text{blue}}{In^-(aq)} \qquad K_a = 1.0 \times 10^{-6}$$

If this indicator is introduced into a very acidic solution, the position of the above equilibrium will be far to the left and the solution will be red. For example, assume that pH = 1.0 ($[H^+] = 1.0 \times 10^{-1}$ $M$):

$$K_a = 1.0 \times 10^{-6} = \frac{[H^+][In^-]}{[HIn]}$$

Rearranging,

$$\frac{[K_a]}{[H^+]} = \frac{[In^-]}{[HIn]}$$

$$\frac{K_a}{[H^+]} = \frac{1.0 \times 10^{-6}}{1.0 \times 10^{-1}} = 10^{-5} = \frac{[In^-]}{[HIn]} = \frac{1}{100,000}$$

HIn is by far the predominant form of the indicator; the solution will be red.

If $OH^-$ is added, the $[H^+]$ concentration decreases, and the equilibrium

$$HIn \rightleftarrows H^+ + In^-$$

shifts to the right (HIn is changed to $In^-$). At some point enough $In^-$ will be present in solution so that a bluish tinge will be noticeable. That is, a color change from red to reddish-purple will occur.

Another way to think about this process is that the stronger acid reacts with the $OH^-$ first; when this acid is gone, the $OH^-$ reacts with the indicator, changing it from HIn to $In^-$, producing the color change. This way of thinking about the process is somewhat oversimplified but may be helpful.

At what pH will the color change be apparent? The normal assumption about indicators is that the color change will be visible when enough of the predominant form (HIn) is converted to the minor form ($In^-$) so that

$$\frac{[In^-]}{[HIn]} = \frac{1}{10}$$

*Remember this assumption.* Although the ratio is arbitrary, it serves well enough for most indicators and is important in almost all problems involving indicators.

**Sample Exercise 3.5** An indicator, HIn ($K_a = 1.0 \times 10^{-6}$) where HIn is red and $In^-$ is blue, is placed in a solution of strong acid. The solution is then titrated with a NaOH solution. At what pH will the indicator color change occur?

**Solution**

For the indicator

$$K_a = 1.0 \times 10^{-6} = \frac{[H^+][In^-]}{[HIn]}$$

Assume the color change is visible when $\frac{[In^-]}{[HIn]} = \frac{1}{10}$

$$K_a = 1.0 \times 10^{-6} = [H^+]\frac{1}{10}$$

$$[H^+] = 1.0 \times 10^{-5}$$

$$pH = 5.0$$

The color change occurs at pH = 5.0.

Note that the color change is not gradual but sharp. This is because the pH changes dramatically near the equivalence point of a titration (see the pH curves in Figures 3.1 and 3.2). Thus in a given titration the [HIn] may be much greater than the $[In^-]$ at one point, but the next drop of $OH^-$ will change the pH so much that the $[In^-]$ may be comparable to or greater than the [HIn]. This drop of $OH^-$ solution, then, will produce the color change. This phenomenon will be discussed more thoroughly in the next section.

**TEST 3.5** Two drops of an indicator, HIn ($K_a = 1.0 \times 10^{-8}$) where HIn is yellow and $In^-$ is blue, are placed in 100.0 mL of 0.30 *M* HCl.

**A.** What color is the solution initially?

**B.** This solution is titrated with 0.20 *M* NaOH. At what pH will the color change (yellow to greenish-yellow) occur?
**C.** What color will the solution be after 300.0 mL of NaOH has been added?

## 3.6 Selection of Indicators for Particular Titrations

The purpose of a titration is usually to answer questions such as: What is the concentration of acid or base in a given solution? What is the molecular weight of an unknown acid? To answer questions of this type, one must know when the equivalence point of the titration is reached. There are two typical methods for determining the equivalence point:

1. Plotting a pH curve by using a pH meter to monitor the pH. The vertical region of the pH curve indicates the equivalence point.
2. Using an indicator. The point where the indicator changes color is called the *endpoint*. The goal of the person doing the titration is to have the endpoint and the equivalence point be identical (or at least within acceptable error limits). The task of choosing an indicator is made easier by the dramatic change of the pH near the equivalence point, as seen from the curves in Figure 3.3.

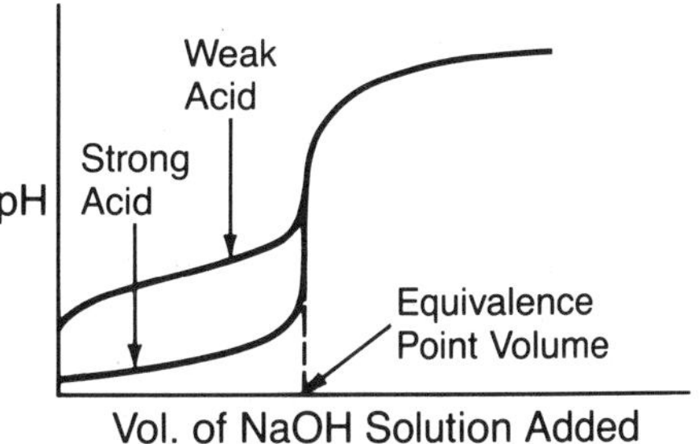


**Figure 3.3**

This dramatic change in pH near the equivalence point produces a sharp endpoint (the color change usually occurs with one drop of added titrant) and minimizes the error if the endpoint and the equivalence point are not identical.

**Sample Exercise 3.6**

Consider the titration of 100.00 mL of 0.100 *M* HCl with 0.1000 *M* NaOH. What should be the $K_a$ value of the indicator used for this titration?

**Solution**

The first question to be answered is: At what pH does the equivalence point occur? Solutions containing $Na^+$, $OH^-$, $H^+$, and $Cl^-$ are mixed in this titration. The titration reaction will be

$$H^+(aq) + OH^-(aq) \longrightarrow H_2O(l)$$

and at the equivalence point the solution will contain the major species

$$Na^+, Cl^-, H_2O$$

Since neither $Na^+$ nor $Cl^-$ is a significant acid or base, the solution will be neutral (pH = 7.00) at the equivalence point. Thus an indicator that changes color at pH 7 will be needed.

Recall the principles discussed in the previous section. The indicator (HIn) will be predominantly in the HIn form in the initially acidic solution. As $OH^-$ is added, the pH will increase rather slowly at first and then will rise dramatically at the equivalence point. This will cause the indicator equilibrium

$$HIn(aq) \rightleftarrows H^+(aq) + In^-(aq)$$

to shift suddenly to the right, producing enough $In^-$ to cause a color change. The color change for a typical indicator can be seen when

$$\frac{[In^-]}{[HIn]} \approx \frac{1}{10}$$

For an indicator that changes color at a pH of 7 ($[H^+] = 1.0 \times 10^{-7}$ *M*)

$$K_a = \frac{[H^+][In^-]}{[HIn]} = [H^+]\left(\frac{[In^-]}{[HIn]}\right)$$

$$K_a = (1.0 \times 10^{-7})\left(\frac{1}{10}\right) = 1.0 \times 10^{-8}$$

Thus an indicator with $K_a = 1.0 \times 10^{-8}$ will change color at approximately pH 7 when an acid is titrated with a base. If, for example, HIn is red and $In^-$ is blue, the color change will be from red to reddish-purple (red with a little blue mixed in).

Another significant question is, How much leeway is there in choosing an indicator? That is, how close do the endpoint of the indicator and the equivalence point of the titration have to be? Consider again the titration of sample exercise 3.6. What is the pH of the solution after 99.99 mL of 0.1000 *M* NaOH have been added? (This is just before the equivalence point.)

$$\text{no. of mmol } H^+ \text{ left} = 10.000 - 9.999 = 0.001 \text{ mmol } H^+$$

$$[H^+] = \frac{0.001 \text{ mmol}}{(100.00 + 99.99) \text{ mL}} = 5 \times 10^{-6}\ M$$

$$\text{pH} = 5.3$$

What is the pH of the solution after 100.01 mL of 0.1000 *M* NaOH has been added? (This is just after the equivalence point, and the $OH^-$ is in excess.)

$$\text{no. of mmol } OH^- \text{ in excess} = 10.001 - 10.000 = 0.001 \text{ mmol } OH^-$$

$$[OH^-] = \frac{0.001 \text{ mmol}}{(100.00 + 100.01) \text{ mL}} = 5 \times 10^{-6}\ M$$

$$[H^+][OH^-] = 1.00 \times 10^{-14} = [H^+](5 \times 10^{-6})$$

$$[H^+] = 2 \times 10^{-9}$$

$$\text{pH} = 8.7$$

These calculations show that in going from 99.99 mL to 100.01 mL of added NaOH solution, the pH will change from 5.3 to 8.7. The difference between 99.99 mL and 100.01 mL is 0.02 mL, which is about half of an ordinary drop. Two important conclusions can be drawn from this phenomenon:

1. Indicator color changes will be sharp. The large change in pH around the equivalence point will cause an appropriate indicator to pass, with the addition of one drop of base, from a condition where the [HIn] is much greater than the $[In^-]$ (the solution being colored by HIn) to a condition where the $[In^-]$ is significant (the solution is now colored by both HIn and $In^-$). Thus the color change occurs with the addition of a single drop of titrant.

2. There is some leeway in choosing the indicator. Consider this titration to be carried out twice using two indicators:
   (a) Indicator A—color change at a pH of 5.
   (b) Indicator B—color change at a pH of 9.
   The results of these titrations would agree within one drop of titrant, which is acceptable error in a typical titration.

The point is that this titration could be done accurately using a wide range of indicators. This is typical of a strong acid–strong base titration.

---

**TEST 3.6** What range of indicator $K_a$ values could be used in the titration of 100.0 mL of 0.10 $M$ HCl with 0.10 $M$ NaOH (assuming that the color change must occur between pH 5 and pH 9)?

---

Now consider the case of a weak acid being titrated with a solution of NaOH. Note from Figure 3.3 that the vertical region around the equivalence point is much smaller for the titration of a weak acid than for a strong acid. This means that there is less leeway in choosing the indicator for the titration of a weak acid than for a strong acid.

**Sample Exercise 3.7** Consider the titration of 50.00 mL of 0.500 $M$ HOCl ($K_a = 3.50 \times 10^{-8}$) with 0.100 $M$ NaOH. What value of $K_a$ should an indicator have to be used to mark the equivalence point of this titration?

**Solution**

The question that must be answered first is: What is the pH at the equivalence point? The titration reaction is

$$\mathrm{HOCl}(aq) + \mathrm{OH^-}(aq) \longrightarrow \mathrm{H_2O}(l) + \mathrm{OCl^-}(aq)$$

Initially the solution contains

$$(50.00 \text{ mL})\,(0.500\ M) = 25.0 \text{ mmol of HOCl}$$

Thus 25.0 mmol of $OH^-$ must be added to reach the equivalence point. What volume of NaOH solution is required?

$$(x \text{ mL})\,(0.100\ M) = 25.0 \text{ mmol}$$

$$x = \frac{25.0}{0.100} = 250.0 \text{ mL}$$

The total volume at the equivalence point is 50.0 mL + 250.0 mL = 300.0 mL. At the equivalence point the major species in solution are $Na^+$, $OCl^-$, $H_2O$.

In considering these species, one concludes that $OCl^-$ is a base (anion of the weak acid, HOCl) and will react with $H_2O$ to produce a basic solution:

$$\mathrm{OCl^-} + \mathrm{H_2O} \rightleftarrows \mathrm{HOCl} + \mathrm{OH^-}$$

$$K_b = \frac{[\mathrm{HOCl}][\mathrm{OH^-}]}{[\mathrm{OCl^-}]} = \frac{K_w}{K_a} = \frac{1.00 \times 10^{-14}}{3.50 \times 10^{-8}} = 2.86 \times 10^{-7}$$

*Initial Concentrations*

$$[\mathrm{OCl^-}]_0 = \frac{25.0 \text{ mmol}}{300.0 \text{ mL}} = 8.33 \times 10^{-2}\ M$$

$\xrightarrow{\text{Let } x \text{ mol/L of } OCl^- \text{ react with } H_2O}$

*Equilibrium Concentrations*

$$[\mathrm{OCl^-}] = 8.33 \times 10^{-2} - x$$

$[HOCl]_0 = 0$ $\qquad$ $[HOCl] = x$

$[OH^-]_0 \approx 0$ $\qquad$ $[OH^-] = x$

$$2.86 \times 10^{-7} = K_b = \frac{[HOCl][OH^-]}{[OCl^-]} = \frac{(x)(x)}{8.33 \times 10^{-2} - x} \approx \frac{x^2}{8.33 \times 10^{-2}}$$

$x^2 \approx 2.38 \times 10^{-8}$

$x \approx 1.54 \times 10^{-4}$

This simplification is valid, so

$$x = 1.54 \times 10^{-4} = [OH^-]$$

$$[H^+][OH^-] = 1.00 \times 10^{-14} = (1.54 \times 10^{-4})\,[H^+]$$

$$[H^+] = 6.49 \times 10^{-11}$$

$$pH = 10.19$$

The pH at the equivalence point is 10.19, so an indicator that changes color near pH 10.19 must be chosen. As usual, the indicator will be a weak acid:

$$HIn \rightleftarrows H^+ + In^-$$

and the color change will be apparent when $[In^-]/[HIn] \approx 1/10$, so

$$K_a = [H^+]\left(\frac{[In^-]}{[HIn]}\right)$$

$$K_a = (6.49 \times 10^{-11})\left(\frac{1}{10}\right) = 6.49 \times 10^{-12}$$

Thus an indicator with a $K_a$ value in the range of $10^{-11}$ to $10^{-12}$ is needed to mark accurately the equivalence point of this titration.

**TEST 3.7** What $K_a$ value should the indicator have to mark the equivalence point of the titration of 100.0 mL of 0.100 *M* HCN ($K_a = 6.2 \times 10^{-10}$) with 0.100 *M* NaOH?

So far in the discussion of indicators, only the titrations of acids with NaOH have been considered. How is the situation different when a base is titrated with a strong acid?

Consider the titration of a base with a solution of HCl. The indicator is a weak acid:

$$\underset{\text{red}}{HIn(aq)} \rightleftarrows H^+(aq) + \underset{\text{blue}}{In^-(aq)}$$

The initially basic solution will be blue, because the $[H^+]$ is very small. This means the equilibrium is shifted far right, and $In^-$ is the predominant form of the indicator. As acid is added during the titration, the pH will decrease slowly until near the equivalence point, when it will decrease rapidly. This will cause a color change as the amount of HIn present increases dramatically. When will the color change be visible? As before, assume that the color change will be apparent when approximately one-tenth of the major (predominant) form is converted to the minor form. In this case, the predominant form is $In^-$ since the solution is basic. Thus when $[In^-]/[HIn]$ is approximately 10/1 the color change will be apparent. Note that the ratio $[In^-]/[HIn]$ for a color change is 1/10 for titration of an acid and 10/1 for titration of a base.

**Sample Exercise 3.8**

An indicator, HIn, has a $K_a$ value of $1.0 \times 10^{-7}$. Determine at what pH the color change occurs when the indicator is used to mark the equivalence point for

**A.** The titration of an acid.
**B.** The titration of a base.

**Solution**

**A.** The solution is initially acidic, so that HIn is the predominant form. The color change will occur when $[In^-]/[HIn] = 1/10$.

$$K_a = 1.0 \times 10^{-7} = \frac{[H^+][In^-]}{[HIn]} = [H^+]\left(\frac{1}{10}\right)$$

$$[H^+]_{\text{color change}} = (10)(1.0 \times 10^{-7}) = 1.0 \times 10^{-6}$$

$$pH = 6.0$$

**B.** The solution here is initially basic and $In^-$ is predominant. The color change will occur when $[In^-]/[HIn] = 10/1$.

$$K_a = 1.0 \times 10^{-7} = \frac{[H^+][In^-]}{[HIn]} = [H^+]\left(\frac{10}{1}\right)$$

$$[H^+]_{\text{color change}} = \frac{1.0 \times 10^{-7}}{10} = 1.0 \times 10^{-8}$$

$$pH = 8.0$$

To summarize, consider the following hypothetical indicator:

$$\underset{\text{red}}{HIn(aq)} \rightleftarrows H^+(aq) + \underset{\text{blue}}{In^-(aq)} \qquad K_a = 1.0 \times 10^{-7}$$

$$K_a = \frac{[H^+][In^-]}{[HIn]} = [H^+]\left(\frac{[In^-]}{[HIn]}\right) = 1.0 \times 10^{-7}$$

| *Titration* | *Color Change* | *Ratio Where Color Change Occurs* | *$[H^+]$ Where Color Change Occurs* | *pH Where Color Change Occurs* |
|---|---|---|---|---|
| Acid with a base | red ⟶ reddish-purple | $\frac{[In^-]}{[HIn]} \approx \frac{1}{10}$ | $1.0 \times 10^{-6}$ | 6.0 |
| Base with an acid | blue ⟶ bluish-purple | $\frac{[In^-]}{[HIn]} \approx \frac{10}{1}$ | $1.0 \times 10^{-8}$ | 8.0 |

Note that when this indicator is used in the titration of a base by an acid, the color change (blue–bluish-purple) occurs at pH 8, whereas when this indicator is used in the titration of an acid with a base, the color change (red–reddish-purple) occurs at pH 6.

**TEST 3.8** The titration of ammonia with a solution of hydrochloric acid produces a pH of about 5 at the equivalence point. What $K_a$ should an indicator have to mark the equivalence point accurately?

## EXERCISES

1. In each of the following cases:
   1. Write the major species before any reaction occurs.
   2. Write the equation for the reaction that occurs.
   3. Write the major species in solution after the reaction.
   4. Determine what equilibrium will control the $[H^+]$.

   (a) 500.0 mL of 0.10 $M$ HCl is mixed with 500.0 mL of 0.20 $M$ NaOH.
   (b) 250.0 mL of 0.50 $M$ NaOH is added to 300.0 mL of 1.0 $M$ HOAc ($K_a = 1.8 \times 10^{-5}$).
   (c) 0.010 mol of HCl($g$) is dissolved in 1.0 L of 0.050 $M$ $NH_3$ ($K_b = 1.8 \times 10^{-5}$). Assume no volume change.
   (d) 500.0 mL of 0.10 $M$ HCl is added to 500.0 mL of 0.20 $M$ NaCN ($K_a$ for HCN is $6.2 \times 10^{-10}$).
   (e) 200.0 mL of 0.30 $M$ NaOH is added to 800.0 mL of 0.10 $M$ $HCHO_2$ (formic acid, $K_a = 1.8 \times 10^{-4}$).
   (f) 500.0 mL of 1.0 $M$ NaOH is added to 500.0 mL of 0.10 $M$ benzoic acid ($C_6H_5COOH$, $K_a = 6.3 \times 10^{-5}$).
   (g) 0.010 mol of $HNO_3$ is added to 1.0 L of 0.10 $M$ $C_5H_5N$ (pyridine, $K_b = 1.4 \times 10^{-9}$). Assume no volume change.
   (h) A solution is formed by mixing 500.0 mL of 0.20 $M$ $NH_3$ ($K_b = 1.8 \times 10^{-5}$) and 500.0 mL of 0.10 $M$ HCl.

2. Calculate the pH of each solution in problem 1.

3. Consider the titration of 75.0 mL of 0.20 $M$ HCl with 0.10 $M$ NaOH. Calculate the pH after the following total volumes of NaOH solution have been added:
   (a) 0.0 mL (b) 50.0 mL
   (c) 75.0 mL (d) 150.0 mL
   (e) 300.0 mL

4. Consider the titration of 25.0 mL of 0.10 $M$ $Ba(OH)_2$ with 0.050 $M$ $HNO_3$. Calculate the pH of the solution after the following total volumes of $HNO_3$ have been added:
   (a) 0.0 mL (b) 10.0 mL
   (c) 25.0 mL (d) 50.0 mL
   (e) 100.0 mL (f) 200.0 mL

5. Consider the titration of 75.0 mL of 0.20 $M$ HCN ($K_a = 6.2 \times 10^{-10}$) with 0.10 $M$ NaOH.
   (a) Calculate the pH after the following total volumes of 0.10 $M$ NaOH have been added:
   1. 0.0 mL 2. 50.0 mL
   3. 75.0 mL 4. 150.0 mL
   5. 300.0 mL

   (b) What $K_a$ should an indicator have to mark correctly the equivalence point of this titration?

6. An indicator HIn has a $K_a$ value of $5.0 \times 10^{-5}$. The equilibrium is

$$\underset{\text{yellow}}{HIn(aq)} \rightleftarrows H^+(aq) + \underset{\text{red}}{In^-(aq)}$$

   (a) Assume this indicator is used to mark the equivalence point for an acid titrated with a base.
   1. What color will the solution be at the beginning of the titration?
   2. At what pH will the color change (endpoint) be apparent?
   3. What is the color change at the endpoint?

(b) Assume this indicator is used to mark the equivalence point of the titration of a base with a strong acid.
   1. What is the color of the solution at the beginning of the titration?
   2. At what pH does the endpoint occur?
   3. What is the color change at the endpoint?

(c) At what pH will this indicator appear medium orange?

**7.** Consider the titration of 25.0 mL of 0.20 $M$ $NH_3$ ($K_b = 1.8 \times 10^{-5}$) with 0.050 $M$ $HNO_3$. Calculate the pH of the solution after the following total volumes of $HNO_3$ have been added:

(a) 0.0 mL (b) 10.0 mL
(c) 25.0 mL (d) 50.0 mL
(e) 100.0 mL (f) 200.0 mL

**Verbalizing General Concepts**

Answer the following in your own words.

**8.** Define titration, equivalence or stoichiometric point, and endpoint.

**9.** Describe the differences between the pH curves for the titration of a strong acid with a sodium hydroxide solution and a weak acid with a sodium hydroxide solution.

**10.** In the titration of a weak acid with sodium hydroxide, why is the pH at the equivalence point greater than 7?

**11.** Explain how an indicator works.

**12.** Why does one have to be more selective in choosing an indicator for the titration of a weak acid than for the titration of a strong acid?

**Multiple Choice Questions**

**13.** A 50.0 mL solution of 1.50 $M$ NaOH is being titrated with a 2.00 $M$ HCl solution. What will the pH be after the addition of 35.0 mL of HCl?

(a) 1.23 (b) 11.7
(c) 12.8 (d) 2.30
(e) 10.5

**14.** In question 13, what will be the final volume of solution when the NaOH has been completely neutralized by the HCl?

(a) 87.5 mL (b) 100.0 mL
(c) 90.0 mL (d) 97.5 mL
(e) 89.0 mL

**15.** Given:

| *Indicator* | *Acid Color* | *pH Range of Color Change* | *Base Color* |
|---|---|---|---|
| thymol blue | yellow | 8.0–9.6 | blue |
| thymol yellow | red | 1.2–2.8 | yellow |
| phenolphthalein | colorless | 8.3–10.0 | red |
| bromcresol green | yellow | 3.8–5.5 | blue |

You are given an unknown solution and asked to determine its basicity or acidity. On adding thymol yellow to half of the solution, it turns yellow, while the other half remains colorless on addition of phenolphthalein. The possible pH range of the unknown is:

(a) 2.8–8.3 (b) 0.0–8.3
(c) 2.8–9.6 (d) 8.0–8.3
(e) 2.8–5.5

For exercises 16 and 17 consider the titration of 100.0 mL of a 0.5000 $M$ HCl solution with 0.2500 $M$ KOH solution.

**16.** What will be the pH after the addition of 150.0 mL of the KOH solution?
(a) 1.08 (b) 1.30
(c) 5.70 (d) 12.7
(e) None of these

**17.** What will be the final volume of solution when the HCl solution has been completely neutralized by the KOH solution?
(a) 150.0 mL (b) 200.0 mL
(c) 250.0 mL (d) 300.0 mL
(e) 400.0 mL

**18.** What is the pH of a $5.00 \times 10^{-3}$ $M$ aqueous solution of a soluble compound with formula $M(OH)_2$, assuming complete dissociation?
(a) 11.7 (b) 2.30
(c) 7.00 (d) 2.00
(e) 12.0

**19.** When 20.0 mL of 0.200 $M$ NaOH is added to 100.0 mL of 0.050 $M$ acetic acid ($HC_2H_3O_2$, $K_a = 1.8 \times 10^{-5}$), the pH of the resulting solution is
(a) 10.7 (b) 5.35
(c) 4.14 (d) 8.32
(e) None of these

**20.** The pH of a solution that results from the addition of 26.0 mL of 0.200 $M$ NaOH to 50.0 mL of 0.100 $M$ HCl is
(a) 2.58 (b) 11.4
(c) 3.42 (d) 10.6
(e) None of these

**21.** When 100.0 mL of 0.500 $M$ NaOH is added to 150.0 mL of a solution of 0.300 $M$ acetic acid ($K_a = 1.8 \times 10^{-5}$), the pH of the resulting solution at equilibrium is
(a) 2.30 (b) 11.70
(c) 12.70 (d) 12.30
(e) 1.70

For exercises 22–24, consider the titration of 200.0 mL of a 0.5000 $M$ HCl solution with 1.000 $M$ NaOH solution.

**22.** What will the pH be after the addition of 75.00 mL of the NaOH solution?
(a) 0.477 (b) 0.903
(c) 1.041 (d) 1.601
(e) None of these

**23.** What will be the final volume of solution when the HCl solution has been completely neutralized by the NaOH solution?
(a) 100.0 mL (b) 275.0 mL
(c) 300.0 mL (d) 400.0 mL
(e) 600.0 mL

**24.** What will be the pH of the solution after the addition of 125.0 mL of the NaOH solution?
(a) 1.114 (b) 8.114
(c) 12.886 (d) 13.114
(e) None of these

**25.** A strong acid is being titrated with a 0.500 $M$ NaOH solution. Which statement is true for this titration?
(a) The pH at the equivalence point cannot be determined without knowing the identity of the acid.
(b) The pH at the equivalence point cannot be determined unless the concentration of the acid is known.

(c) The pH at the equivalence point will be less than 7.0, since an acid is being titrated.
(d) The pH at the equivalence point will be 7.0.
(e) The pH at the equivalence point will be greater than 7.0, since a strong base is being used.

For exercises 26–28, consider the titration of 100.0 mL of 0.500 $M$ $NH_3$ with 0.500 $M$ HCl. The $K_b$ for $NH_3$ is $1.8 \times 10^{-5}$.

**26.** After 50.0 mL of HCl has been added, the $[H^+]$ of the solution is
(a) $1.8 \times 10^{-5}$ $M$
(b) $5.6 \times 10^{-10}$ $M$
(c) $1.2 \times 10^{-5}$ $M$
(d) $1.0 \times 10^{-7}$ $M$
(e) None of these

**27.** How many mL of 0.500 $M$ HCl are required to reach the stoichiometric point of this reaction?
(a) 25.0 mL
(b) 50.0 mL
(c) 100.0 mL
(d) 200.0 mL
(e) None of these

**28.** At the stoichiometric point of this titration, the $[H^+]$ is
(a) $1.0 \times 10^{-7}$ $M$
(b) $1.8 \times 10^{-5}$ $M$
(c) $1.2 \times 10^{-5}$ $M$
(d) $5.6 \times 10^{-10}$ $M$
(e) None of these

**29.** 50.0 mL of 0.10 $M$ $HNO_2$ ($K_a = 4.0 \times 10^{-4}$) is being titrated with 0.10 $M$ NaOH. The pH after 25.0 mL of NaOH has been added is
(a) 7.00
(b) 1.00
(c) 12.50
(d) 3.40
(e) None of these

**30.** In the titration of a weak acid, HA, with 0.100 $M$ NaOH, the stoichiometric point is known to occur at a pH value of approximately 10. Which of the following indicator acids would be best to use to mark the endpoint of this titration?
(a) Indicator A, $K_a = 10^{-14}$
(b) Indicator B, $K_a = 10^{-11}$
(c) Indicator C, $K_a = 10^{-8}$
(d) Indicator D, $K_a = 10^{-6}$
(e) None of these will work well

**31.** Consider the titration of 100.0 mL of 0.100 $M$ NaOH with 1.00 $M$ HCl. How much 1.00 $M$ HCl must be added to reach a pH of 12.0?
(a) 10.0 mL
(b) 9.52 mL
(c) 10.5 mL
(d) 8.91 mL
(e) None of these

For exercises 32–39 consider the titration of 60.0 mL of 0.250 $M$ formic acid ($HCHO_2$, $K_a = 1.8 \times 10^{-4}$) with 0.200 $M$ NaOH.

**32.** The pH before any NaOH is added is
(a) 4.35
(b) 2.17
(c) 0.60
(d) 3.74
(e) None of these

**33.** The pH after 15.0 mL of 0.200 $M$ NaOH has been added is
(a) 4.34
(b) 3.74
(c) 0.80
(d) 3.14
(e) None of these

**34.** The pH after 25.0 mL of 0.200 $M$ NaOH has been added is
(a) 3.44
(b) 4.05
(c) 3.74
(d) 0.88
(e) None of these

35. The pH after 37.5 mL of 0.200 $M$ NaOH has been added is
(a) 1.11 (b) 1.87
(c) 3.74 (d) 7.0
(e) None of these

36. The pH after 60.0 mL of 0.200 $M$ NaOH has been added is
(a) 4.35 (b) 3.14
(c) 1.60 (d) 3.74
(e) None of these

37. The pH after 75.0 mL of 0.200 $M$ NaOH has been added is
(a) 7.00 (b) 5.60
(c) 8.40 (d) 9.46
(e) None of these

38. The pH after 100.0 mL of 0.200 $M$ NaOH has been added is
(a) 1.51 (b) 13.10
(c) 14.00 (d) 12.49
(e) None of these

39. The indicator most appropriate to mark the equivalence point of this titration would have a $K_a$ value of
(a) $4 \times 10^{-10}$ (b) $4 \times 10^{-5}$
(c) $4 \times 10^{-13}$ (d) $4 \times 10^{-8}$
(e) None of these

40. A 50.0 mL solution of 1.50 $M$ NaOH is being titrated with a 3.00 $M$ HCl solution. What will the pH be after the addition of 35.0 mL of HCl?
(a) 1.23 (b) 11.70
(c) 0.45 (d) 2.30
(e) 10.50

41. In exercise 40, what will be the final volume of solution when the NaOH has been completely neutralized by the HCl?
(a) 75.0 mL (b) 100.0 mL
(c) 90.0 mL (d) 97.5 mL
(e) 89.0 mL

42. What volume of 0.0100 $M$ NaOH must be added to 1.00 L of 0.0500 $M$ HA ($K_a = 4.0 \times 10^{-8}$) to achieve a pH of 8.00?
(a) 1.00 L (b) 5.00 L
(c) 1.00 L (d) 4.00 L
(e) None of these

43. What is the $[OH^-]$ after 50.00 mL of 0.100 $M$ $HNO_3$ and 50.10 mL of 0.100 $M$ NaOH have been mixed?
(a) $1.0 \times 10^{-4}$ $M$ (b) $2.0 \times 10^{-12}$ $M$
(c) $1.0 \times 10^{-10}$ $M$ (d) $5.0 \times 10^{-3}$ $M$
(e) None of these

For exercises 44–51, consider the titration of 100.0 mL of 0.100 $M$ methylamine ($CH_3NH_2$, $K_b = 4.4 \times 10^{-4}$) with 0.100 $M$ HCl.

44. Calculate the pH before any HCl has been added.
(a) 2.18 (b) 11.82
(c) 13.00 (d) 10.64
(e) None of these

45. Calculate the pH after 10.0 mL of 0.100 $M$ HCl has been added.
(a) 11.60 (b) 2.40
(c) 4.31 (d) 9.69
(e) None of these

46. Calculate the pH after 25.0 mL of 0.100 $M$ HCl has been added.
(a) 2.88 (b) 11.12
(c) 3.83 (d) 10.17
(e) None of these

47. Calculate the pH after 50.0 mL of 0.100 $M$ HCl has been added.
(a) 7.00 (b) 1.48
(c) 3.36 (d) 10.64
(e) None of these

48. Calculate the pH after 75.0 mL of 0.100 $M$ HCl has been added.
(a) 2.88 (b) 11.12
(c) 3.83 (d) 10.17
(e) None of these

49. Calculate the pH after 100.0 mL of 0.100 $M$ HCl has been added.
(a) 9.03 (b) 5.97
(c) 7.00 (d) 3.36
(e) None of these

50. Calculate the pH after 200.0 mL of 0.100 $M$ HCl has been added.
(a) 0.30 (b) 1.00
(c) 0.18 (d) 1.48
(e) None of these

51. The indicator most appropriate to mark the equivalence point of this titration would have the $K_a$ value
(a) $1 \times 10^{-5}$ (b) $1 \times 10^{-7}$
(c) $1 \times 10^{-3}$ (d) $1 \times 10^{-10}$
(e) None of these

For exercises 52 and 53, consider the titration of 100.0 mL of 0.250 $M$ aniline with 0.500 $M$ HCl. The $K_b$ for aniline is $3.8 \times 10^{-10}$.

52. What is the pH of the solution at the stoichiometric point?
(a) −0.78 (b) 5.10
(c) 2.68 (d) 11.32
(e) None of these

53. In the calculation of the volume of HCl required to reach a pH of 8.0, which of the following expressions is correct? ($x$ = volume of HCl (in mL) required to reach a pH of 8.0.)
(a) $\dfrac{0.5x - (100)(0.25)}{100 + x} = [\text{aniline}]$ (b) $[\text{H}^+] = x$
(c) $\dfrac{0.5x}{100 + x} = [\text{aniline}]$ (d) $\dfrac{25 - 0.5x}{100 + x} - 10^{-6} = [\text{aniline}]$
(e) None of these

54. 0.0100 mol of a weak acid (HA) was dissolved in 100.0 mL of distilled water, and a pH titration was performed using 0.100 $M$ NaOH. After 40.00 mL of 0.100 $M$ NaOH were added, the pH of the solution was observed to be 4.00. The $K_a$ value for HA is
(a) $1.0 \times 10^{-4}$ (b) $6.7 \times 10^{-5}$
(c) $1.5 \times 10^{-4}$ (d) $1.0 \times 10^{-10}$
(e) None of these

# 4 Solubility Equilibria

## CHAPTER OBJECTIVES

1. Define solubility, solubility product constant, solubility product expression, ion product, and precipitation.
2. Calculate the $K_{sp}$ value for a solid, given the solubility of the solid.
3. Calculate the solubility of a solid in pure water, given the $K_{sp}$ of the solid.
4. Calculate the solubility of a solid (given its $K_{sp}$) in a solution containing a common ion.
5. Know the principles involved in predicting relative solubilities of solids, given their $K_{sp}$ values.
6. Know the principles involved in predicting whether precipitation will occur when two solutions are mixed.
7. Know the principles involved in the separation of cations by selective precipitation.

## 4.1 Introduction

When a typical ionic solid is dissolved in water, the dissolved material can be assumed to be present as separate hydrated anions and cations. For example, consider $CaF_2(s)$ dissolving in water. The process can be represented as

$$CaF_2(s) \xrightarrow{H_2O} Ca^{2+}(aq) + 2F^-(aq)$$

The solubility of a substance is defined in terms of the amount of material that will dissolve in a given amount of solvent (for example, grams of solute per 100 mL of solvent) or the amount of material that will dissolve to produce a certain volume of the resulting solution (for example, moles of solute per liter of solution). The latter system will be used here to define solubility.

Consider the process that occurs as $CaF_2(s)$ is added to water. Initially the water contains no $Ca^{2+}$ or $F^-$ ions, but as the solid dissolves, the concentrations of $Ca^{2+}$ and $F^-$ build up in solution. As the concentrations of $Ca^{2+}$ and $F^-$ increase, it becomes more and more likely that these ions will collide and reform the solid phase. Thus competing processes are occurring:

$$CaF_2(s) \longrightarrow Ca^{2+}(aq) + 2F^-(aq) \tag{4.1}$$

and the reverse reaction

$$Ca^{2+}(aq) + 2F^-(aq) \longrightarrow CaF_2(s) \tag{4.2}$$

At first, the dissolving process, (4.1), dominates and solid disappears. As the concentrations of $Ca^{2+}$ and $F^-$ build up, however, the rate of the reverse process, (4.2), increases until it equals the rate of dissolving. At this point, dynamic equilibrium is reached: the solid is dissolving and reforming at the same rate, and no net change occurs in the amount of remaining undissolved solid. The combined process can be represented as

$$CaF_2(s) \rightleftarrows Ca^{2+}(aq) + 2F^-(aq)$$

and an equilibrium expression can be constructed according to the Law of Mass Action:

$$K_{sp} = [Ca^{2+}][F^-]^2$$

$CaF_2(s)$ is not included, since pure condensed phases (solids and liquids) are never included in the equilibrium expression. This means that the amount of excess solid does not affect the position of the solubility equilibrium and thus does not affect the solubility of that solid.

This observation might seem strange at first. One might think that the more solid present, and thus the more surface area exposed to the solvent (water), the greater the solubility. However, when the ions in solution reform the solid, they do so at the surface of the solid. Doubling the surface area of the solid not only doubles the rate of dissolving but also doubles the rate of reformation of the solid, so that the amount of excess solid present has no effect on the equilibrium position.

Similarly, while grinding up the solid (to increase the surface area) or stirring the solution speeds up the attainment of equilibrium, neither procedure changes the *amount* of solid dissolved at equilibrium.

In summary, note the following points:

1. The equilibrium constant is given a special symbol ($K_{sp}$) and is called the *solubility product constant* ($K_{\text{solubility product}}$).
2. Solubility is an *equilibrium position.* The solubility of a substance and the solubility constant for a substance are *not* the same thing.
3. The solid itself is not included in the equilibrium expression.

## 4.2 Solving Typical Solubility Equilibrium Problems

The most common solubility problems fall into two categories:

1. Calculation of the value of $K_{sp}$ from the solubility of a solid.
2. Calculation of the solubility of a solid, given the $K_{sp}$ value.

Both of these situations will be considered in sample exercises.

**Sample Exercise 4.1** The solubility of $CuBr(s)$ is $2.0 \times 10^{-4}$ mol/L at 25°C. Calculate the value of the $K_{sp}$ for $CuBr(s)$ at 25°C.

**Solution**

When $CuBr(s)$ is placed in water, it dissolves by producing $Cu^+$ and $Br^-$ ions according to the following equilibrium:

$$CuBr(s) \rightleftarrows Cu^+(aq) + Br^-(aq) \qquad K_{sp} = [Cu^+][Br^-]$$

Now follow the procedure of defining initial and equilibrium concentrations as in acid-base problems:

*Initial Concentrations*
*(before the equilibrium of interest occurs)*

$$[Cu^+]_0 = 0$$
$$[Br^-]_0 = 0$$

Now let the system come to equilibrium by dissolving CuBr(*s*). Let $x$ = mol/L of CuBr(*s*) that are dissolved when the system comes to equilibrium. That is, $x$ mol/L CuBr(*s*) dissolve to produce $x$ mol/L $Cu^+$ plus $x$ mol/L $Br^-$ in solution. At equilibrium,

$$[Cu^+] = x$$
$$[Br^-] = x$$

The solubility has been given: Solubility = $2.0 \times 10^{-4}$ mol/L, which means that $2.0 \times 10^{-4}$ mol of CuBr(*s*) dissolves to produce $2.0 \times 10^{-4}$ mol of $Cu^+$ and $2.0 \times 10^{-4}$ mol of $Br^-$ per liter of the resulting solution.

$$K_{sp} = [Cu^+][Br^-] = (x)(x) = (2.0 \times 10^{-4})(2.0 \times 10^{-4})$$
$$= 4.0 \times 10^{-8}\ (\text{mol/L})^2$$

This equilibrium constant is characteristic of CuBr(*s*) in water at 25°C. When excess CuBr(*s*) is placed in contact with an aqueous solution of CuBr(*s*) at 25°C, the product of $[Cu^+]$ and $[Br^-]$ must always be equal to $4.0 \times 10^{-8}\ (\text{mol/L})^2$.

Note the following points:

1. When a $K_{sp}$ value is given, the units are usually deleted.
2. In calculating the $K_{sp}$ value of CuBr(*s*) from the measured solubility, it has been assumed that all of the CuBr(*s*) that dissolves does so to form separate $Cu^+$ and $Br^-$ ions. Although some of the ions may exist as ion pairs ($Cu^+ \cdots Br^-$) in solution, this complication will be ignored.
3. If the solubility is given in units of g/100 mL, it must be changed to mol/L before doing equilibrium calculations.

**TEST 4.1** The solubility of silver iodide (AgI) is $1.2 \times 10^{-8}$ mol/L at 25°C. Calculate the $K_{sp}$ for AgI(*s*) at 25°C.

**A.** What equilibrium occurs when AgI(*s*) is placed in water?
**B.** Write the $K_{sp}$ expression for AgI(*s*).
**C.** Determine the initial concentrations.
**D.** Define $x$.
**E.** Write the equilibrium concentrations in terms of $x$.
**F.** Determine the value of $x$.
**G.** Calculate the $K_{sp}$ value.

**Sample Exercise 4.2** The solubility of $CaF_2(s)$ is $2.15 \times 10^{-4}$ mol/L at 25°C. Calculate the $K_{sp}$ for $CaF_2(s)$ at 25°C.

**Solution**

The usual steps will be done but will be represented in condensed form. The equilibrium of interest is

$$CaF_2(s) \rightleftarrows Ca^{2+}(aq) + 2F^-(aq) \qquad K_{sp} = [Ca^{2+}][F^-]^2$$

| *Initial Concentrations* | | *Equilibrium Concentrations* |
|---|---|---|
| $[Ca^{2+}]_0 = 0$ | Let $x$ mol/L of $CaF_2(s)$ dissolve $\longrightarrow$ | $[Ca^{2+}] = x$ |
| $[F^-]_0 = 0$ | | $[F^-] = 2x$ |

Note: $xCaF_2 \longrightarrow xCa^{2+} + 2xF^-$
Solubility $= x = 2.15 \times 10^{-4}$ mol/L

$$[Ca^{2+}] = x = 2.15 \times 10^{-4} \text{ mol/L}$$

$$[F^-] = 2x = 2(2.15 \times 10^{-4} \text{ mol/L}) = 4.30 \times 10^{-4} \text{ mol/L}$$

$$K_{sp} = [Ca^{2+}][F^-]^2 = (2.15 \times 10^{-4})(4.30 \times 10^{-4})^2$$

$$K_{sp} = 3.98 \times 10^{-11}$$

**TEST 4.2** The solubility of $Bi_2S_3(s)$ is $1.0 \times 10^{-15}$ mol/L at 25°C. Calculate the $K_{sp}$ for $Bi_2S_3$ at 25°C. Remember to go through the steps described above.

In the preceding exercises, the $K_{sp}$ value has been calculated from known solubility. In the following exercises, solubility will be calculated from the $K_{sp}$.

**Sample Exercise 4.3** The $K_{sp}$ for $Cu(IO_3)_2(s)$ is $1.4 \times 10^{-7}$ at 25°C. Calculate the solubility of $Cu(IO_3)_2(s)$ at 25°C.

**Solution**

The solubility equilibrium is

$$Cu(IO_3)_2(s) \rightleftarrows Cu^{2+}(aq) + 2IO_3^-(aq) \qquad K_{sp} = [Cu^{2+}][IO_3^-]^2$$

| *Initial Concentrations (before any $Cu(IO_3)_2(s)$ dissolves)* | | *Equilibrium Concentrations* |
|---|---|---|
| $[Cu^{2+}]_0 = 0$ | Let $x$ mol/L $Cu(IO_3)_2(s)$ dissolve $\longrightarrow$ | $[Cu^{2+}] = x$ |
| $[IO_3^-]_0 = 0$ | | $[IO_3^-] = 2x$ |

Note: $xCu(IO_3)_2(s) \longrightarrow xCu^{2+} + 2xIO_3^-$

$$1.4 \times 10^{-7} = K_{sp} = [Cu^{2+}][IO_3^-]^2 = (x)(2x)^2 = 4x^3$$

$$\frac{1.4 \times 10^{-7}}{4} = 3.5 \times 10^{-8} = x^3$$

$$x = \sqrt[3]{3.5 \times 10^{-8}} = \sqrt[3]{35} \times 10^{-3} = 3.3 \times 10^{-3} \text{ mol/L}$$

$x =$ mol/L of $Cu(IO_3)_2$ that dissolves = solubility

Solubility $= 3.3 \times 10^{-3}$ mol/L

**Sample Exercise 4.4** The $K_{sp}$ value for AgBr($s$) is $7.7 \times 10^{-13}$ at 25°C. Calculate the solubility of AgBr($s$) in water at 25°C.

**Solution**

The equilibrium is

$$AgBr(s) \rightleftarrows Ag^+(aq) + Br^-(aq)$$

and the equilibrium expression is

$$K_{sp} = [Ag^+][Br^-] = 7.7 \times 10^{-13}$$

| *Initial Concentrations (before any AgBr(s) dissolves)* | | *Equilibrium Concentrations* |
|---|---|---|
| $[Ag^+]_0 = 0$ | Let $x$ mol/L of AgBr($s$) dissolve ⟶ | $[Ag^+] = x$ |
| $[Br^-]_0 = 0$ | | $[Br^-] = x$ |

Note: $x AgBr(s) \longrightarrow x Ag^+ + x Br^-$

$$7.7 \times 10^{-13} = K_{sp} = [Ag^+][Br^-] = (x)(x)$$

$$x^2 = 7.7 \times 10^{-13}$$

$$x = 8.8 \times 10^{-7} \text{ mol/L} = \text{solubility}$$

**TEST 4.3** The $K_{sp}$ for silver chromate ($Ag_2CrO_4$) is $9.0 \times 10^{-12}$ at 25°C. Calculate the solubility of $Ag_2CrO_4$ at 25°C.

**A.** Write the solubility equilibrium.
**B.** Write the $K_{sp}$ expression.
**C.** Write down the initial concentrations.
**D.** Define $x$.
**E.** Write the equilibrium concentrations in terms of $x$.
**F.** Substitute the equilibrium concentrations into the $K_{sp}$ expression.
**G.** Calculate $x$.
**H.** Find the solubility.

**Sample Exercise 4.5** Rank the following solids in order of decreasing solubility.

**A.** AgI($s$) ($K_{sp} = 1.5 \times 10^{-16}$), AgBr($s$) ($K_{sp} = 7.7 \times 10^{-13}$), AgCl($s$) ($K_{sp} = 1.6 \times 10^{-10}$), $SrSO_4$($s$) ($K_{sp} = 2.9 \times 10^{-7}$), CuI($s$) ($K_{sp} = 5.0 \times 10^{-12}$), and $CaSO_4$($s$) ($K_{sp} = 6.1 \times 10^{-5}$)

**Solution**

Each of these solids dissolves to produce two ions:

$$MA(s) \rightleftarrows M^{n+}(aq) + A^{n-}(aq) \qquad K_{sp} = [M^{n+}][A^{n-}]$$

If $x$ = solubility, then in each case at equilibrium

$$[M^{n+}] = x$$

$$[A^{n-}] = x$$

$$K_{sp} = [M^{n+}][A^{n-}] = (x)(x)$$

$$x^2 = K_{sp}$$

$$x = \sqrt{K_{sp}} = \text{solubility}$$

Thus the order of solubility for these solids can be determined by ordering the $K_{sp}$ values: the solid with the largest $K_{sp}$ has the highest solubility. The order is

$$CaSO_4(s) > SrSO_4(s) > AgCl(s) > CuI(s) > AgBr(s) > AgI(s)$$

most soluble　　　　least soluble

**B.** $CuS(s)$ ($K_{sp} = 8.5 \times 10^{-45}$), $Ag_2S(s)$ ($K_{sp} = 1.6 \times 10^{-49}$), and $Bi_2S_3(s)$ ($K_{sp} = 1.1 \times 10^{-73}$).

**Solution**

In this case, each solid produces a different number of ions so that the $K_{sp}$ values *cannot* be compared directly to determine the relative solubilities. Calculate the solubility for each solid as a test of your knowledge (**Test 4.4**).

## 4.3 Common Ion Effect

In the cases so far considered, the solid has been dissolved in pure water. Problems in which the solution contains an ion in common with the one in the salt will now be introduced.

**Sample Exercise 4.6** Calculate the solubility $Ag_2CrO_4(s)$ in a 0.100 $M$ solution of $AgNO_3$. ($K_{sp}$ for $Ag_2CrO_4$ is $9.0 \times 10^{-12}$.)

**Solution**

Before any $Ag_2CrO_4$ dissolves, the solution contains $Ag^+$, $NO_3^-$, and $H_2O$. The $Ag_2CrO_4(s)$ dissolves as follows:

$$Ag_2CrO_4(s) \rightleftarrows 2Ag^+(aq) + CrO_4^{2-}(aq)$$

$$K_{sp} = [Ag^+]^2[CrO_4^{2-}] = 9.0 \times 10^{-12}$$

*Initial Concentrations (before any $Ag_2CrO_4(s)$ dissolves)*

$[Ag^+]_0 = 0.100\ M$ from the dissolved $AgNO_3$

$[CrO_4^{2-}]_0 = 0$

Let $x$ mol/L of $Ag_2CrO_4$ dissolve. This produces $xCrO_4^{2-}$ and $2xAg^+$.

*Equilibrium Concentrations*

$[Ag^+]_0 = 0.100 + 2x$

$[CrO_4^{2-}]_0 = x$

$9.0 \times 10^{-12} = K_{sp} = [Ag^+]^2[CrO_4^{2-}] = (0.100 + 2x)^2(x)$

Since the $K_{sp}$ for $Ag_2CrO_4$ is small (the position of the equilibrium lies far to the left), $x$ is expected to be small compared to 0.100 $M$. Therefore, assume that $0.100 + 2x \approx 0.100$ and

$$9.0 \times 10^{-12} = (0.100 + 2x)^2x \approx (0.100)^2x$$

$$x \approx \frac{9.0 \times 10^{-12}}{(0.100)^2} = 9.0 \times 10^{-10}\ \text{mol/L}$$

Note that $x$ is much less than 0.100 $M$, so that the assumption is valid.

$$\text{Solubility} = x = 9.0 \times 10^{-10} \text{ mol/L}$$

The equilibrium concentrations are

$$[Ag^+] = 0.100 + 2(9.0 \times 10^{-10}) = 0.100\ M$$

$$[CrO_4^{2-}] = x = 9.0 \times 10^{-10} \text{ mol/L}$$

Now compare the solubility of $Ag_2CrO_4(s)$ in pure water (calculated in Test 4.3) and in 0.100 $M$ $AgNO_3$.

| | *Pure Water* | *0.100 M $AgNO_3$* |
|---|---|---|
| Solubility of $Ag_2CrO_4$ | $1.3 \times 10^{-4}$ mol/L | $9.0 \times 10^{-10}$ mol/L |

Notice that the solubility of $Ag_2CrO_4(s)$ is much less when the water contains $Ag^+$ from $AgNO_3$. This is an example of the *common ion effect*: The presence in solution of ions in common with one or more of those in the solid being dissolved lowers the solubility of the solid.

**TEST 4.5** Calculate the solubility of $Ag_3PO_4(s)$ ($K_{sp} = 1.8 \times 10^{-18}$) in pure water and in a 1.0 $M$ $Na_3PO_4$ solution and compare the results.

## 4.4 Precipitation Conditions

When solutions are mixed, reactions often occur. Acid-base reactions were considered in chapter 3. In this section, reactions that cause precipitation (formation of a solid) will be considered.

The *ion product*, another name for a reaction quotient, will be used to solve these problems. The ion product expression is identical to the equilibrium expression for a given solid except that *initial concentrations* are used. This differs from the $K_{sp}$ expression, which involves *equilibrium concentrations* only. For example, for the compound $CaF_2(s)$, the ion product expression is

$$\text{Ion product} = Q = [Ca^{2+}]_0[F^-]_0^2$$

When a solution containing $Ca^{2+}$ is added to a solution containing $F^-$, precipitation (formation of $CaF_2(s)$) may or may not occur. Precipitation *will occur* if $Q > K_{sp}$.

To predict whether precipitation of a given solid will occur, calculate the ion product for the solid using the concentrations of the ions in solution. Then apply the following rules:

If $Q > K_{sp}$ precipitation occurs

If $Q \leqslant K_{sp}$ no precipitation occurs

**Sample Exercise 4.7** A solution is prepared by mixing 100.0 mL of $1.0 \times 10^{-3}$ $M$ $Ca(NO_3)_2$ and 100.0 mL of $1.0 \times 10^{-3}$ $M$ NaF. Does $CaF_2(s)$ precipitate from this solution? ($K_{sp}$ for $CaF_2(s)$ is $4.0 \times 10^{-11}$.)

**Solution**

Since $Ca(NO_3)_2(s)$ and $NaF(s)$ are soluble salts, the $Ca(NO_3)_2$ solution contains $Ca^{2+}$ and $NO_3^-$ and the NaF solution contains $Na^+$ and $F^-$. To see if $CaF_2$ forms, the first thing to do is to compute the concentrations of $Ca^{2+}$ and $F^-$ in the mixed solution (which has a volume of 200.0 mL):

$$[Ca^{2+}]_0 = \frac{\text{no. of mmol } Ca^{2+}}{\text{mL of solution}} = \frac{(100.0 \text{ mL})(1.0 \times 10^{-3}\ M)}{200.0 \text{ mL}}$$

$$= 5.0 \times 10^{-4}\ M$$

$$[F^-]_0 = \frac{\text{no. of mmol } F^-}{\text{mL of solution}} = \frac{(100.0 \text{ mL})(1.0 \times 10^{-3}\ M)}{200.0 \text{ mL}}$$

$$= 5.0 \times 10^{-4}\ M$$

When $CaF_2(s)$ dissolves, the reaction is

$$CaF_2(s) \rightleftarrows Ca^{2+}(aq) + 2F^-(aq)$$

and the ion product is

$$Q = [Ca^{2+}]_0[F^-]_0^2$$

In this case,

$$Q = [Ca^{2+}]_0[F^-]_0^2 = (5.0 \times 10^{-4})(5.0 \times 10^{-4})^2 = 1.25 \times 10^{-10}$$

$$K_{sp} = 4.0 \times 10^{-11}$$

Thus $Q > K_{sp}$ and $CaF_2(s)$ will form.

**TEST 4.6** A mixture is formed by adding 50.0 mL of $1.0 \times 10^{-2}\ M$ $CuNO_3$ to 200.0 mL of $1.0 \times 10^{-4}\ M$ NaCl. Does $CuCl(s)$ ($K_{sp} = 1.8 \times 10^{-7}$) form?

## 4.5 Selective Precipitation

Precipitation is sometimes used to separate the ions in a mixture. For example, consider a solution containing $Ba^{2+}$, $Ag^+$, and $NO_3^-$. The $Ba^{2+}$ and the $Ag^+$ can be separated in at least two different ways by precipitating one of the ions while leaving the other in solution. If NaCl is added to the $Ba^{2+}$, $Ag^+$ mixture, $AgCl(s)$ (a white solid) will precipitate. Since $BaCl_2$ is soluble, it will not form, and $Ba^{2+}$ will be left in solution.

Another way to accomplish this separation is to add $Na_2SO_4$ to the $Ag^+$, $Ba^{2+}$ mixture. $BaSO_4(s)$ will precipitate, leaving $Ag^+$ in solution ($Ag_2SO_4(s)$ is very soluble).

Separation of ions by precipitation is called *selective precipitation.*

Since metal sulfides differ dramatically in solubility, sulfide ion is often used to separate metal ions. This type of separation will be considered in the following example.

**Sample Exercise 4.8** A solution contains $1.0 \times 10^{-3}\ M$ $Fe(NO_3)_2$ and $1.0 \times 10^{-3}\ M$ $MnSO_4$. Both $Fe^{2+}$ and $Mn^{2+}$ form sulfide salts: FeS ($K_{sp} = 3.7 \times 10^{-19}$) and MnS ($K_{sp} = 1.4 \times 10^{-15}$). If sulfide ion is added to the solution containing $10^{-3}\ M$ $Fe^{2+}$ and $10^{-3}\ M$ $Mn^{2+}$, which sulfide salt will precipitate first and at what $[S^{2-}]$ will precipitation occur?

**Solution**

The solution contains $1.0 \times 10^{-3}$ $M$ $Fe^{2+}$ and $1.0 \times 10^{-3}$ $M$ $Mn^{2+}$. Precipitation of each sulfide will occur when the ion product is greater than the $K_{sp}$ for that salt.

For FeS ($FeS(s) \rightleftarrows Fe^{2+}(aq) + S^{2-}(aq)$), the ion product is

$$Q = [Fe^{2+}]_0[S^{2-}]_0$$

Since $[Fe^{2+}]_0 = 1.0 \times 10^{-3}$ $M$,

$$Q = (1.0 \times 10^{-3})[S^{2-}]_0$$

$K_{sp}$ for FeS is $3.7 \times 10^{-19}$. If $Q = K_{sp}$, no precipitation occurs, but any added $S^{2-}$ will cause precipitation.

Let $Q = K_{sp} = 3.7 \times 10^{-19} = (1.0 \times 10^{-3})[S^{2-}]_0$. Then

$$[S^{2-}]_0 = \frac{3.7 \times 10^{-19}}{1.0 \times 10^{-3}} = 3.7 \times 10^{-16}\ M$$

Thus when $[S^{2-}] = 3.7 \times 10^{-16}$ $M$, no precipitation of FeS($s$) will occur, but if $[S^{2-}] > 3.7 \times 10^{-16}$ $M$ in this solution, FeS($s$) will form.

Now do the same calculations for MnS:

$$Q = [Mn^{2+}]_0[S^{2-}]_0$$

$$[Mn^{2+}]_0 = 1.0 \times 10^{-3}\ M$$

Let $Q = K_{sp} = 1.4 \times 10^{-15} = (1.0 \times 10^{-3})[S^{2-}]_0$

$$[S^{2-}]_0 = \frac{1.4 \times 10^{-15}}{1.0 \times 10^{-3}} = 1.4 \times 10^{-12}\ M$$

If $[S^{2-}]_0 > 1.4 \times 10^{-12}$ $M$, MnS($s$) will precipitate.

The questions posed above can now be answered. As $S^{2-}$ is added to the solution containing $10^{-3}$ $M$ $Fe^{2+}$ and $10^{-3}$ $M$ $Mn^{2+}$, the FeS($s$) will precipitate first (when $[S^{2-}] > 3.7 \times 10^{-16}$ $M$) and the MnS($s$) will precipitate when $[S^{2-}] > 1.4 \times 10^{-12}$ $M$.

---

**TEST 4.7** A solution contains a mixture of $1.0 \times 10^{-2}$ $M$ $Pb(NO_3)_2$ and $1.0 \times 10^{-4}$ $M$ $AgNO_3$. The $K_{sp}$ values for $PbCl_2$ and AgCl are $1.0 \times 10^{-4}$ and $1.6 \times 10^{-10}$, respectively. As $Cl^-$ is added to this solution, which salt will precipitate first and at what concentration of $Cl^-$?

---

## 4.6 Selective Precipitation of Sulfides

Sulfide forms many salts with widely varying solubilities. These salts are often used to precipitate metal ions selectively from solution in qualitative analysis schemes.

The key to the selective precipitation of sulfide salts is the basicity of $S^{2-}$, which allows its concentration to be controlled by controlling the pH of the solution.

Sulfide forms the acid $H_2S$, for which the following equilibrium expression can be written

$$\frac{[H^+]^2[S^{2-}]}{[H_2S]} = K = 1.32 \times 10^{-20}$$

To form sulfide precipitates, the solution is saturated with $H_2S$, which produces a 0.10 $M$ concentration of $H_2S$. Thus, for a saturated solution, the equilibrium expression can be written

$$\frac{[H^+]^2[S^{2-}]}{[H_2S]} = \frac{[H^+][S^{2-}]}{(0.10)} = 1.32 \times 10^{-20}$$

$$[H^+][S^{2-}] = (0.10)(1.32 \times 10^{-20}) = 1.3 \times 10^{-21}$$

Note from this expression that the concentration of $S^{2-}$ can be regulated by controlling the $[H^+]$. A large $[H^+]$ means a small $[S^{2-}]$ and vice versa.

**Sample Exercise 4.9** Calculate the $[S^{2-}]$ in a solution saturated with $H_2S$ where the pH is 1.00.

**Solution**

The pH = 1.00. To find the $[H^+]$, take the antilog of pH:

$$[H^+] = 10^{-pH} = 1.0 \times 10^{-1}\ M$$

The equilibrium expression for $H_2S$ in a saturated solution is

$$[H^+]^2[S^{2-}] = 1.3 \times 10^{-21}$$

$$[S^{2-}] = \frac{1.3 \times 10^{-21}}{[H^+]^2} = \frac{1.3 \times 10^{-21}}{(1.0 \times 10^{-1})^2}$$

$$[S^{2-}] = 1.3 \times 10^{-19}\ M$$

**TEST 4.8** Calculate the $[S^{2-}]$ in a solution saturated with $H_2S$ where the pH is 9.00.

The results of sample exercise 4.9 and test 4.8 show that the $[S^{2-}]$ can be varied over many orders of magnitude by varying the pH. This allows metal sulfide salts to be precipitated selectively, as will be shown in the following example.

**Sample Exercise 4.10** Consider a solution that contains $1.0 \times 10^{-3}\ M\ Mn^{2+}$ and $1.0 \times 10^{-3}\ M\ Cu^{2+}$, is saturated with $H_2S$, and has a pH of 2.00. Under these conditions, does either $CuS(s)$ ($K_{sp} = 8.5 \times 10^{-45}$) or $MnS(s)$ ($K_{sp} = 2.3 \times 10^{-13}$) form?

**Solution**

The solution is saturated with $H_2S$, so the relationship

$$[H^+]^2[S^{2-}] = 1.3 \times 10^{-21}$$

can be used. The pH = 2.00, which means that $[H^+]$ is obtained from the antilog of pH:

$$[H^+] = 10^{-pH} = 1.0 \times 10^{-2}\ M$$

$$[S^{2-}] = \frac{1.3 \times 10^{-21}}{[H^+]^2} = \frac{1.3 \times 10^{-21}}{(1.0 \times 10^{-2})^2} = 1.3 \times 10^{-17}\ M$$

Now compute $Q$ for each salt to see if precipitation occurs.

For MnS($s$): $Q = [Mn^{2+}]_0[S^{2-}]_0 = (1.0 \times 10^{-3})(1.3 \times 10^{-17}) = 1.3 \times 10^{-20}$
$K_{sp} = 2.3 \times 10^{-13}$
$Q < K_{sp}$, so no MnS($s$) forms.

For CuS($s$): $Q = [Cu^{2+}]_0[S^{2-}]_0 = (1.0 \times 10^{-3})(1.3 \times 10^{-17}) = 1.3 \times 10^{-20}$
$K_{sp} = 8.5 \times 10^{-45}$
$Q > K_{sp}$, so CuS($s$) forms.

Under these conditions, $Cu^{2+}$ precipitates as CuS($s$), while $Mn^{2+}$ remains in solution. Thus a separation of $Cu^{2-}$ from $Mn^{2+}$ has been achieved. At pH 2 the $[S^{2-}]$ is large enough to precipitate CuS($s$) but not the more soluble MnS($s$).

**TEST 4.9** Consider a solution that contains $1.0 \times 10^{-2}$ $M$ $Fe^{2+}$ and $1.0 \times 10^{-2}$ $M$ $Ni^{2+}$. The solution is saturated with $H_2S$ and has a pH of 1.00.

Under these conditions, will NiS($s$) ($K_{sp} = 1.4 \times 10^{-24}$) or FeS($s$) ($K_{sp} = 3.7 \times 10^{-19}$) precipitate?

## EXERCISES

1. The solubility of CuI($s$) in water is $2.3 \times 10^{-6}$ mol/L. Calculate the $K_{sp}$ for CuI($s$).
2. The solubility of $PbI_2(s)$ is $1.2 \times 10^{-3}$ mol/L. Calculate the $K_{sp}$ for $PbI_2(s)$.
3. The $K_{sp}$ for $ZnC_2O_4(s)$ is $1.5 \times 10^{-9}$. Calculate the solubility of $ZnC_2O_4(s)$.
4. The $K_{sp}$ for $SrF_2(s)$ is $2.8 \times 10^{-9}$. Calculate the solubility of $SrF_2(s)$.
5. The solubility of $Ca_3(PO_4)_2(s)$ is $6.3 \times 10^{-7}$ mol/L. Calculate the $K_{sp}$ value for $Ca_3(PO_4)_2(s)$.
6. The $K_{sp}$ for AgBr($s$) is $7.7 \times 10^{-13}$. Calculate the solubility of AgBr($s$) in
   (a) Pure water.
   (b) $1.0 \times 10^{-2}$ $M$ NaBr solution.
7. A solution is saturated with $BaSO_4(s)$ (solid is added until an excess remains). The $[Ba^{2+}]$ in this solution is $1.0 \times 10^{-5}$ $M$. Calculate the value of $K_{sp}$ for $BaSO_4(s)$.
8. The $K_{sp}$ for $Pb_3(PO_4)_2(s)$ is $1.0 \times 10^{-42}$. Calculate the solubility of $Pb_3(PO_4)_2(s)$ in
   (a) Pure water.
   (b) 0.100 $M$ $Pb(NO_3)_2$.
9. Calculate the solubility of $Co(OH)_2(s)$ ($K_{sp} = 5.2 \times 10^{-15}$) in a solution with a pH of 11.00.
10. For each of the following pairs of solids, determine which solid is least soluble.
    (a) $CaF_2(s)$ ($K_{sp} = 4.0 \times 10^{-11}$) or $BaF_2(s)$ ($K_{sp} = 1.1 \times 10^{-6}$)
    (b) $Ca_3(PO_4)_2(s)$ ($K_{sp} = 1.0 \times 10^{-29}$) or $FePO_4(s)$ ($K_{sp} = 1.0 \times 10^{-22}$)

11. A solution contains $1.0 \times 10^{-5}$ $M$ $Ag^+$ and $2.0 \times 10^{-6}$ $M$ $CN^-$. Will $AgCN(s)$ precipitate? ($K_{sp}$ for $AgCN(s)$ is $2.2 \times 10^{-12}$.)

12. A solution contains $2.0 \times 10^{-3}$ $M$ $Ce^{3+}$ and $1.0 \times 10^{-2}$ $M$ $IO^{3-}$. Will $Ce(IO_3)_3(s)$ precipitate? (The $K_{sp}$ for $Ce(IO_3)_3$ is $3.2 \times 10^{-10}$.)

13. A solution contains $1.0 \times 10^{-8}$ $M$ $Co^{2+}$. At what $[S^{2-}]$ will precipitation of $CoS(s)$ ($K_{sp} = 7.0 \times 10^{-23}$) begin?

14. A solution is prepared by mixing 50.0 mL of $1.0 \times 10^{-3}$ $M$ $Ca(NO_3)_2$ and 50.0 mL of $1.0 \times 10^{-2}$ $M$ $Na_2SO_4$. Will $CaSO_4(s)$ ($K_{sp} = 6.1 \times 10^{-5}$) precipitate?

15. A solution is prepared by mixing 75.0 mL of $1.0 \times 10^{-3}$ $M$ $CuNO_3$ and 150.0 mL of $1.0 \times 10^{-3}$ $M$ NaCl. Will $CuCl(s)$ ($K_{sp} = 1.8 \times 10^{-7}$) precipitate?

16. A solution is prepared by mixing 100.0 mL of $1.0 \times 10^{-2}$ $M$ $Pb(NO_3)_2$ and 100.0 mL of $1.0 \times 10^{-3}$ $M$ NaF. Will $PbF_2(s)$ ($K_{sp} = 3.7 \times 10^{-8}$) precipitate?

17. A solution contains $1.0 \times 10^{-5}$ $M$ $PO_4^{3-}$. What is the minimum concentration of $Ag^+$ that would cause precipitation of $Ag_3PO_4(s)$ ($K_{sp} = 1.8 \times 10^{-18}$)?

18. The solubility of $Pb(IO_3)_2(s)$ in a 0.10 $M$ $KIO_3$ solution is $2.6 \times 10^{-11}$ mol/L. Calculate the $K_{sp}$ for $Pb(IO_3)_2(s)$.

## Verbalizing General Concepts

Answer the following in your own words.

19. Define solubility (in two types of units), $K_{sp}$, ion product, selective precipitation, and the common ion effect.

20. Discuss the fact that solubility is an equilibrium position.

21. Under what conditions can $K_{sp}$ values be compared directly to determine the relative solubilities of salts?

22. How can one determine whether a precipitate will form when two solutions are mixed?

## Multiple Choice Questions

23. The $[IO_3^-]$ in a solution in equilibrium with $Ce(IO_3)_3(s)$ is $5.55 \times 10^{-3}$ $M$. Calculate the $K_{sp}$ for $Ce(IO_3)_3(s)$.
(a) $3.16 \times 10^{-10}$ (b) $1.03 \times 10^{-5}$
(c) $2.56 \times 10^{-8}$ (d) $3.51 \times 10^{-11}$
(e) None of these

24. A solution is $1 \times 10^{-4}$ $M$ in each of $F^-$, $S^{2-}$, and $PO_4^{3-}$. What would be the order of precipitation as $Pb^{2+}$ is added?
$K_{sp}(PbF_2) = 4 \times 10^{-8}$, $K_{sp}(PbS) = 3 \times 10^{-28}$, $K_{sp}(Pb_3(PO_4)_2) = 1 \times 10^{-42}$
(a) $PbF_2$ before PbS before $Pb_3(PO_4)_2$.
(b) $Pb_3(PO_4)_2$ before PbS before $PbF_2$.
(c) PbS before $PbF_2$ before $Pb_3(PO_4)_2$.
(d) $Pb_3(PO_4)_2$ before $PbF_2$ before PbS.
(e) PbS before $Pb_3(PO_4)_2$ before $PbF_2$.

25. The concentration of $Ag^+$ in a saturated solution of $Ag_2C_2O_4(s)$ is $2.2 \times 10^{-4}$ $M$. The $K_{sp}$ of $Ag_2C_2O_4(s)$ is
(a) $5.3 \times 10^{-12}$ (b) $4.8 \times 10^{-8}$
(c) $1.1 \times 10^{-11}$ (d) $2.2 \times 10^{-4}$
(e) None of these

**26.** The $K_{sp}$ of a metal sulfide, $MS(s)$, is $2.0 \times 10^{-17}$. The sulfide ion concentration of a solution containing $MS(s)$ at equilibrium is
(a) $4.5 \times 10^{-9}$ *M*
(b) $1.0 \times 10^{-17}$ *M*
(c) $2.0 \times 10^{-17}$ *M*
(d) $2.3 \times 10^{-9}$ *M*
(e) None of these

**27.** How many moles of $CaF_2$ will dissolve in 1.0 L of a 0.025 *M* NaF solution? ($K_{sp}$ for $CaF_2 = 4.0 \times 10^{-11}$)
(a) $6.4 \times 10^{-8}$
(b) $1.6 \times 10^{-9}$
(c) $4.0 \times 10^{-11}$
(d) $2.5 \times 10^{-14}$
(e) None of these

**28.** 0.005 mol $Na_2SO_4$ is added to 500.0 mL of each of two solutions, one containing $1.5 \times 10^{-3}$ *M* $BaCl_2$, the other $1.5 \times 10^{-3}$ *M* $CaCl_2$. Given that $K_{sp}$ for $BaSO_4 = 1.0 \times 10^{-10}$ and $K_{sp}$ for $CaSO_4 = 6.1 \times 10^{-5}$,
(a) $BaSO_4$ would precipitate, but $CaSO_4$ would not.
(b) $CaSO_4$ would precipitate, but $BaSO_4$ would not.
(c) Both $BaSO_4$ and $CaSO_4$ would precipitate.
(d) Neither $BaSO_4$ nor $CaSO_4$ would precipitate.
(e) Not enough information is given to determine if precipitation would occur.

**29.** The solubility of $Mg(OH)_2$ (MW = 58.3) is $8.34 \times 10^{-5}$ g/10.0 mL at 25°C. The value of $K_{sp}$ for $Mg(OH)_2$ at 25°C is
(a) $4.09 \times 10^{-8}$
(b) $5.84 \times 10^{-12}$
(c) $11.7 \times 10^{-12}$
(d) $16.7 \times 10^{-12}$
(e) None of these

**30.** The solubility of $CaF_2(s)$ ($K_{sp} = 4.0 \times 10^{-11}$) in 1.0 L of a $1.0 \times 10^{-2}$ *M* solution of NaF is
(a) $4.0 \times 10^{-9}$ mol
(b) $1.0 \times 10^{-7}$ mol
(c) $7.0 \times 10^{-4}$ mol
(d) $4.0 \times 10^{-7}$ mol
(e) None of these

**31.** Which of the following salts shows the smallest solubility in water?
(a) $Bi_2S_3$ ($K_{sp} = 1 \times 10^{-73}$)
(b) $Ag_2S$ ($K_{sp} = 2 \times 10^{-49}$)
(c) MnS ($K_{sp} = 2 \times 10^{-13}$)
(d) HgS ($K_{sp} = 1 \times 10^{-52}$)
(e) $Mg(OH)_2$ ($K_{sp} = 1 \times 10^{-11}$)

**32.** In a solution prepared by adding excess $PbI_2(s)$ ($K_{sp} = 6.9 \times 10^{-9}$) to water, the $[I^-]$ is
(a) $1.5 \times 10^{-3}$ mol/L
(b) $2.4 \times 10^{-3}$ mol/L
(c) $1.2 \times 10^{-3}$ mol/L
(d) $8.4 \times 10^{-5}$ mol/L
(e) None of these

**33.** The $K_{sp}$ for $CuI(s)$ is $5.3 \times 10^{-12}$. The number of moles of $CuI(s)$ that will dissolve in 1.0 L of $1.0 \times 10^{-2}$ *M* KI is
(a) $2.3 \times 10^{-6}$
(b) $5.3 \times 10^{-12}$
(c) $5.3 \times 10^{-8}$
(d) $5.3 \times 10^{-10}$
(e) None of these

**34.** A solution contains $1.0 \times 10^{-4}$ *M* $Zn^{2+}$, $1.0 \times 10^{-5}$ *M* $Ag^+$, 1.0 *M* $H^+$, and is saturated with $H_2S$ (0.10 *M*). Using the following data,

$$ZnS(s) \qquad K_{sp} = 1 \times 10^{-23}$$

$$Ag_2S(s) \qquad K_{sp} = 2 \times 10^{-49}$$

$$1.3 \times 10^{-20} = \frac{[H^+]^2[S^{2-}]}{[H_2S]}$$

determine which statement is correct:
(a) $ZnS(s)$ will form but not $Ag_2S(s)$.
(b) $Ag_2S(s)$ will form but not $ZnS(s)$.
(c) Both $ZnS(s)$ and $Ag_2S(s)$ will form.
(d) Neither $ZnS(s)$ nor $Ag_2S(s)$ will form.

**35.** Consider a solution containing $10^{-3}$ $M$ $Bi^{3+}$, $10^{-3}$ $M$ $Co^{2+}$, and 0.1 $M$ $H_2S$. Which of the following values for $[H^+]$ would allow precipitation of one of these ions as its sulfide salt while leaving the other ion in solution? The $K_{sp}$ for $Bi_2S_3(s)$ is $1.1 \times 10^{-73}$, and the $K_{sp}$ for $CoS(s)$ is $7.0 \times 10^{-23}$.

$$K = \frac{[H^+]^2[S^{2-}]}{[H_2S]} = 1.3 \times 10^{-20}$$

(a) 4.0 $M$
(b) 15.0 $M$
(c) $1.0 \times 10^{-2}$ $M$
(d) $1.0 \times 10^{-1}$ $M$
(e) None of these

# 5 More Involved Equilibrium Problems

## 5.1 Introduction

In this chapter, some systems will be explored that require more elaborate treatment than those dealt with in the earlier chapters. Equilibrium calculations for the following cases will be considered:

1. Acid-base problems in which the quadratic formula must be used.
2. Acid-base problems in which water makes a significant contribution to the $[H^+]$.
3. Polyprotic acids.
4. Precipitation reactions.
5. Complex ions.
6. Solids dissolving in solutions in which secondary reactions occur.

## 5.2 Use of the Quadratic Formula

The following is an acid-base problem that requires use of the quadratic formula.

**Sample Exercise 5.1** Calculate the pH of a 0.10 *M* solution of $HIO_3$ ($K_a = 1.7 \times 10^{-1}$).

**Solution**

Go through the steps as usual:

1. In solution are $HIO_3$, $H_2O$.
2. Both $HIO_3$ and $H_2O$ are acids, but $HIO_3$ is much stronger ($K_a$ is much greater than $K_w$), and it will dominate in the production of $H^+$.
3. The equilibrium of interest is

$$HIO_3(aq) \rightleftarrows H^+(aq) + IO_3^-(aq) \qquad K_a = \frac{[H^+][IO_3^-]}{[HIO_3]}$$

4. Initial concentrations:

$$[HIO_3]_0 = 0.10M$$
$$[IO_3^-]_0 = 0$$
$$[H^+]_0 \approx 0 \quad \text{(ignore the contribution from } H_2O\text{)}$$

5. Let $x$ = mol/L of $HIO_3$ that dissociate.

6. Equilibrium concentrations:

$$[HIO_3] = 0.10 - x$$

$$[IO_3^-] = x$$

$$[H^+] = x$$

7. Substitute into the $K_a$ expression:

$$1.7 \times 10^{-1} = K_a = \frac{[H^+][IO_3^-]}{[HIO_3]} = \frac{(x)(x)}{0.10 - x} \approx \frac{(x)(x)}{0.10}$$

Try this approximation as usual.

$$x^2 \approx (0.10)(1.7 \times 10^{-1}) = 1.7 \times 10^{-2}$$

$$x \approx 1.3 \times 10^{-1}$$

Note that $x$ is larger than 0.10. This is clearly impossible. Thus the approximation is invalid. In order to solve the quadratic equation obtained above, either the quadratic formula or trial and error must be used.

8. For a general quadratic equation of the form $ax^2 + bx + c = 0$, the solution is given by

$$x = \frac{-b \pm \sqrt{b^2 - 4ac}}{2a}$$

In the present problem,

$$K_a = 1.7 \times 10^{-1} = \frac{x^2}{0.10 - x}$$

Now rearrange this expression so that it is written in the general form for a quadratic equation:

$$(1.7 \times 10^{-1})(0.10 - x) = x^2$$

$$(1.7 \times 10^{-2}) - (1.7 \times 10^{-1})x = x^2$$

$$x^2 + (1.7 \times 10^{-1})x - (1.7 \times 10^{-2}) = 0$$

where

$$a = 1$$

$$b = 1.7 \times 10^{-1}$$

$$c = -1.7 \times 10^{-2}$$

Thus

$$x = \frac{(-1.7 \times 10^{-1}) \pm \sqrt{(1.7 \times 10^{-1})^2 - (4)(1)(-1.7 \times 10^{-2})}}{2(1)}$$

$$= \frac{-1.7 \times 10^{-1} \pm \sqrt{(2.9 \times 10^{-2}) + (6.8 \times 10^{-2})}}{2}$$

$$= \frac{-1.7 \times 10^{-1} \pm \sqrt{9.7 \times 10^{-2}}}{2} = \frac{-1.7 \times 10^{-1} \pm 3.1 \times 10^{-1}}{2}$$

$$= -2.4 \times 10^{-1}, +7.0 \times 10^{-2}$$

Since $x = [H^+]$, the negative root is physically impossible. Thus

$$x = 7.0 \times 10^{-2} \text{ mol/L} = [H^+]$$

$$pH = -\log(7.0 \times 10^{-2}) = 1.15$$

**Sample Exercise 5.2**

A solution is prepared by dissolving 0.050 mol of $Fe(NO_3)_3(s)$ in enough water to make a total volume of 1.0 L. The $K_a$ for $Fe(OH_2)_6^{3+}$ is $6.0 \times 10^{-3}$. Calculate the pH of this solution.

**Solution**

Since $Fe(NO_3)_3(s)$ is an ionic compound, the major species are

$$Fe^{3+}(aq),\ NO_3^-(aq),\ H_2O(l)$$

where $Fe^{3+}(aq)$ is $Fe(OH_2)_6^{3+}$. (This information is conveyed in the description of the problem, since the $K_a$ for $Fe(OH_2)_6^{3+}$ is given.)

Consider the acid-base properties of each component:

1. $Fe(OH_2)_6^{3+}$ Weak acid ($K_a$ given)
2. $NO_3^-$ Extremely weak base (conjugate base of strong acid, $HNO_3$)
3. $H_2O$ Very weak acid or base

Comparing the $K_a$ for $Fe(OH_2)_6^{3+}$ with $K_w$ indicates that $Fe(OH_2)_6^{3+}$ will be dominant.

Remembering the definition of $K_a$,

$$Fe(OH_2)_6^{3+} \rightleftarrows Fe(OH)(OH_2)_5^{2+} + H^+$$

$$K_a = \frac{[Fe(OH)(OH_2)_5^{2+}][H^+]}{[Fe(OH_2)_6^{3+}]} = 6.0 \times 10^{-3}$$

| *Initial Concentrations* | | *Equilibrium Concentrations* |
|---|---|---|
| $[Fe(OH_2)_6^{3+}]_0 = 0.050\ M$ | | $[Fe(OH_2)_6^{3+}] = 0.050 - x$ |
| $[H^+]_0 \approx 0$ (Neglect $H^+$ from $H_2O$) | Let $x$ mol/L $Fe(OH_2)_6^{3+}$ dissociate → | $[H^+] = x$ |
| $[Fe(OH)(OH_2)_5^{2+}]_0 = 0$ | | $[Fe(OH)(OH_2)_5^{2+}] = x$ |

$$K_a = 6.0 \times 10^{-3} = \frac{(x)(x)}{0.050 - x} \approx \frac{x^2}{0.050}$$

↖making the usual approximation

$$x^2 \approx 3.0 \times 10^{-4}$$

$$x \approx 1.7 \times 10^{-2}$$

Check the assumption:

$$\frac{x}{[Fe(OH_2)_6^{3+}]_0}(100) = \frac{1.7 \times 10^{-2}}{5.0 \times 10^{-2}} \times 100 = 34\%$$

Thus the approximation $[HA]_0 - x \approx [HA]_0$ is *not* valid. To calculate the correct answer, use the quadratic formula:

$$\frac{x^2}{0.050 - x} = 6.0 \times 10^{-3}$$

$$x^2 = (3.0 \times 10^{-4}) - (6.0 \times 10^{-3})x$$

$$x^2 + (6.0 \times 10^{-3})x - (3.0 \times 10^{-4}) = 0$$

which has the form $ax^2 + bx + c = 0$, where $a = 1$, $b = 6.0 \times 10^{-3}$, and $c = -3.0 \times 10^{-4}$. Then

$$x = \frac{-b \pm \sqrt{b^2 - 4ac}}{2a} = \frac{-6.0 \times 10^{-3} \pm \sqrt{(6.0 \times 10^{-3})^2 - (4)(1)(-3.0 \times 10^{-4})}}{(2)(1)}$$

Disregarding the negative root,

$$x = 1.5 \times 10^{-2} = [H^+]$$

$$pH = 1.82$$

**TEST 5.1** Calculate the pH of $2.00 \times 10^{-3}$ $M$ $HNO_2$ ($K_a = 4.0 \times 10^{-4}$). Be sure to follow the usual steps.

## 5.3 Acid Solutions in Which Water Contributes to [$H^+$]

In all previous calculations involving acids and bases, we have been able to ignore the contribution of water to the total [$H^+$] in solution. This assumption is usually valid, but complications occur when the acid is very weak or very dilute, or both.

### Very Dilute Solutions of Strong Acids

**Sample Exercise 5.3** Calculate the [$H^+$] in a $1.0 \times 10^{-7}$ $M$ HCl solution.

**Solution**

The solution contains the major species

$$H^+, Cl^-, H_2O$$

Normally, the $H^+$ from the strong acid dominates. However, in this case, the strong acid is so dilute that water will make a comparable contribution.

One possible way to approach this problem would be to add the [$H^+$] from the HCl to the [$H^+$] normally found in pure water:

$$[H^+]_{total} \stackrel{?}{=} 1.0 \times 10^{-7} + 1.0 \times 10^{-7} = 2.0 \times 10^{-7}\ M$$

Is this a valid approach?

The answer is no, because the $H^+$ from the strong acid will cause the water equilibrium

$$H_2O(l) \rightleftarrows H^+(aq) + OH^-(aq)$$

to shift to the left. This causes the contribution from the dissociation of $H_2O$ to be less than $1.0 \times 10^{-7}$ $M$. We must use a more elaborate method.

The best way to handle this problem is to recognize that the net charge on this solution must be zero. That is, the positive charge carried by the cations must be exactly equal to the negative charge carried by the anions. This leads to the **charge balance** expression:

$$\text{Concentration of positive charge} = \text{Concentration of negative charge}$$

In the $10^{-7}$ $M$ HCl solution, the only cation is $H^+$ and the only anions are $Cl^-$ and $OH^-$. Thus in this solution

$$[H^+] = [Cl^-] + [OH^-] \qquad \text{charge balance equation}$$

The $Cl^-$ in this solution comes from the dissolved HCl. Thus

$$[Cl^-] = 1.0 \times 10^{-7}\ M$$

$$[H^+] = [Cl^-] + [OH^-] = 1.0 \times 10^{-7}\ M + [OH^-]$$

This equation can be put in terms of one unknown ($[H^+]$) by remembering that

$$[H^+][OH^-] = K_w = 1.0 \times 10^{-14}$$

$$[OH^-] = \frac{1.0 \times 10^{-14}}{[H^+]}$$

Thus

$$[H^+] = [Cl^-] + [OH^-] = [Cl^-] + \frac{1.0 \times 10^{-14}}{[H^+]}$$

$$= 1.0 \times 10^{-7} + \frac{1.0 \times 10^{-14}}{[H^+]}$$

↖ from the HCl

$$[H^+] - \frac{1.0 \times 10^{-14}}{[H^+]} = 1.0 \times 10^{-7}$$

Multiplying both sides of this equation by $[H^+]$ gives

$$[H^+]^2 - (1.0 \times 10^{-14}) = (1.0 \times 10^{-7})[H^+]$$

$$[H^+]^2 - (1.0 \times 10^{-7})[H^+] - (1.0 \times 10^{-14}) = 0$$

This is a quadratic equation in standard form, where $a = 1$, $b = -1.0 \times 10^{-7}$, and $c = -1.0 \times 10^{-14}$. Substitution into the quadratic formula gives

$$[H^+] = \frac{-b \pm \sqrt{b^2 - 4ac}}{2a}$$

$$= \frac{-(-1.0 \times 10^{-7}) \pm \sqrt{(-1.0 \times 10^{-7})^2 - (4)(1)(-1.0 \times 10^{-14})}}{2(1)}$$

$$= 1.6 \times 10^{-7}\ M$$

Thus the $[H^+]$ in a $1.0 \times 10^{-7}$ $M$ HCl solution is $1.6 \times 10^{-7}$ $M$. Since $1.0 \times 10^{-7}$ $M$ $H^+$ comes from the dissolved HCl, $0.6 \times 10^{-7}$ $M$ comes from water. This shows that the position of the water equilibrium,

$$H_2O(l) \rightleftarrows H^+(aq) + OH^-(aq)$$

is shifted to the left by the common ion, $H^+$.

For the $1.0 \times 10^{-7}$ $M$ HCl solution,

$$pH = -\log[H^+] = -\log(1.6 \times 10^{-7}) = 6.80$$

**Sample Exercise 5.4**

Calculate the pH of $1.0 \times 10^{-10}$ $M$ $HNO_3$.

**Solution**

Here is a case where considering only the $H^+$ from the strong acid ($HNO_3$) leads to a clearly ridiculous result. The ionization of $1.0 \times 10^{-10}$ $M$ $HNO_3$ produces $1.0 \times 10^{-10}$ $M$ $H^+$. Then

$$pH = -\log(1.0 \times 10^{-10}) = 10.0$$

This cannot be correct. Adding an acid to water cannot produce a basic solution. Clearly, water must be considered as a source of $H^+$.

To handle this problem, again consider the charge balance equation.

$$[H^+] = [NO_3^-] + [OH^-]$$

$[H^+]$ ↑ from $HNO_3$ and $H_2O$; $[NO_3^-]$ ↖ from $HNO_3$; $[OH^-]$ ↖ from $H_2O$

$$[NO_3^-] = 1.0 \times 10^{-10}\ M$$

Using the $K_w$ expression,

$$[OH^-] = \frac{K_w}{[H^+]} = \frac{1.0 \times 10^{-14}}{[H^+]}$$

Substituting in the above equation gives

$$[H^+] = (1.0 \times 10^{-10}) + \frac{1.0 \times 10^{-14}}{[H^+]}$$

Multiplying both sides by $[H^+]$ gives

$$[H^+]^2 = (1.0 \times 10^{-10})[H^+] + (1.0 \times 10^{-14})$$

$$[H^+]^2 - (1.0 \times 10^{-10})[H^+] - (1.0 \times 10^{-14}) = 0$$

Use of the quadratic formula gives

$$[H^+] = 1.0 \times 10^{-7}\ M$$

$$pH = 7.00$$

Note that the pH is the same as that for pure water within the number of significant figures allowed by the data. This is not really surprising. The amount of $HNO_3$ added is very small compared with the $[H^+]$ from water:

$$10^{-10} \ll 10^{-7}$$

In this solution water is the dominant source of $H^+$.

**TEST 5.2**

**A.** Calculate the $[H^+]$ in a $2.0 \times 10^{-8}$ $M$ solution of $HNO_3$.

**B.** The considerations involved in treating dilute solutions of strong bases are similar to those for strong acids. Try this problem: Calculate the $[OH^-]$ in a $1.0 \times 10^{-7}$ $M$ solution of NaOH.

1. What ions are present?
2. What is the charge balance equation?
3. What is the $[Na^+]$?
4. What is $[H^+]$ in terms of $K_w$ and $[OH^-]$?
5. Solve for $[OH^-]$.

C. The method for calculating the $[H^+]$ using the charge balance equation also gives the correct answer for more concentrated strong acid solutions (the usual case). Show this by using the method to calculate the pH of a 0.10 $M$ HCl solution.

---

## Very Weak Acids or Dilute Solutions of Weak Acids

This situation is more complicated than that of a strong acid, since two equilibria must be considered:

$$H_2O(l) \rightleftarrows H^+(aq) + OH^-(aq) \qquad K_w$$

$$HA(aq) \rightleftarrows H^+(aq) + A^-(aq) \qquad K_a$$

Note that there are four unknowns in this system: $[H^+]$, $[OH^-]$, $[HA]$, and $[A^-]$. To solve for four unknowns requires four equations. The equilibrium equations are two of these. Another useful relationship is the charge balance equation:

$$[H^+] = [A^-] + [OH^-]$$

Knowing that all of the HA originally dissolved must be present at equilibrium either as $A^-$ or HA leads to the equation

$$[HA]_0 = [HA] + [A^-]$$

↖

original concentration of HA dissolved

This is called the **material balance** equation.

These four equations can be used to derive an equation involving only $[H^+]$, as follows:

$$K_a = \frac{[H^+][A^-]}{[HA]}$$

To express $[A^-]$ and $[HA]$ in terms of $[H^+]$, use the charge balance equation

$$[H^+] = [A^-] + [OH^-]$$

where

$$[OH^-] = \frac{K_w}{[H^+]} \quad \text{from the } K_w \text{ expression}$$

The charge balance equation becomes

$$[H^+] = [A^-] + \frac{K_w}{[H^+]}$$

$$[A^-] = [H^+] - \frac{K_w}{[H^+]}$$

This gives $[A^-]$ in terms of $[H^+]$.

The material balance equation is

$$[HA]_0 = [HA] + [A^-] \quad \text{or} \quad [HA] = [HA]_0 - [A^-]$$

Since

$$[A^-] = [H^+] - \frac{K_w}{[H^+]}$$

we have

$$[HA] = [HA]_0 - \left([H^+] - \frac{K_w}{[H^+]}\right)$$

Now substitute into the $K_a$ expression:

$$K_a = \frac{[H^+][A^-]}{[HA]} = \frac{[H^+]\left([H^+] - \frac{K_w}{[H^+]}\right)}{[HA]_0 - \left([H^+] - \frac{K_w}{[H^+]}\right)} = \frac{[H^+]^2 - K_w}{[HA]_0 - \frac{[H^+]^2 - K_w}{[H^+]}}$$

This expression permits the calculation of the $[H^+]$ in a solution containing a weak acid. It gives the correct $[H^+]$ for any solution made by dissolving a weak acid in pure water.

The equation can be solved by simple trial and error or by the more systematic method of successive approximations. The usual way of doing successive approximations is to substitute a guessed value of the variable of interest ($[H^+]$ in this case) into the equation everywhere it appears except one place. The equation is then solved to obtain a value of the variable, which becomes the "guessed value" in the next round. The process is continued until the calculated value equals the guessed value.

Even though the full equation can be solved in this manner, the process is tedious and time-consuming. We would certainly like to use the simpler method (ignoring the contribution of water to the $[H^+]$) wherever possible. Thus, a key question is: Under what conditions can problems involving a weak acid be done the simple way? This question will be answered in the following paragraphs.

Notice that the term $[H^+]^2 - K_w$ appears twice in the full equation:

$$K_a = \frac{[H^+]^2 - K_w}{[HA]_0 - \frac{[H^+]^2 - K_w}{[H^+]}}$$

Under conditions where

$$[H^+]^2 \gg K_w$$

and thus

$$[H^+]^2 - K_w \approx [H^+]^2$$

the full equation reduces as follows:

$$K_a = \frac{[H^+]^2 - K_w}{[HA]_0 - \frac{[H^+]^2 - K_w}{[H^+]}} \approx \frac{[H^+]^2}{[HA]_0 - \frac{[H^+]^2}{[H^+]}} = \frac{[H^+]^2}{[HA]_0 - [H^+]}$$

$$= \frac{x^2}{[HA]_0 - x}$$

where $x = [H^+]$ at equilibrium.

This is an important result: If $[H^+]^2 \gg K_w$, the full equation reduces to the simple expression for a weak acid derived in chapter 2, ignoring water as a source of $H^+$. Let's arbitrarily decide that "much greater than" means at least a factor of 100 times greater. Since $K_w = 1.0 \times 10^{-14}$, the $[H^+]^2$ must be at least $100 \times 10^{-14}$ or $10^{-12}$. This value of $[H^+]^2$ requires that $[H^+] = 10^{-6}$. Thus if $[H^+]$ is greater than or equal to $10^{-6}$ *M*, the complicated equation reduces to the simple equation, i.e., you get the *same answer* by using either equation.

How do we decide when it is necessary to use the complicated equation? The best way to proceed is as follows:

Calculate the $[H^+]$ in the normal way, ignoring $H_2O$. If $[H^+]$ from this calculation is greater than or equal to $10^{-6}$ *M*, the answer is correct, i.e., the complicated equation will give the same answer. If $[H^+]$ is less than $10^{-6}$ *M*, you must use the full equation, i.e., water must be considered as a source of $H^+$.

**Sample Exercise 5.5**

**A.** Calculate the $[H^+]$ in 1.0 $M$ HCN ($K_a = 6.2 \times 10^{-10}$).

**Solution**

First do the weak acid problem the "normal" way. This leads to the expression

$$\frac{x^2}{1.0 - x} = 6.2 \times 10^{-10} \approx \frac{x^2}{1.0}$$

$$x = 2.5 \times 10^{-5}\ M = [H^+]$$

Note that $[H^+]$ is greater than $10^{-6}$ $M$, so we are finished.

**TEST 5.3** Show that $[H^+] = 2.5 \times 10^{-5}$ $M$ also satisfies the full equation (where $H_2O$ is taken into account) for a 1.0 $M$ HCN solution.

**B.** Calculate the $[H^+]$ in $1.0 \times 10^{-4}$ $M$ HCN ($K_a = 6.2 \times 10^{-10}$).

**Solution**

First do the weak acid problem the "normal" way. This leads to the expression

$$K_a = 6.2 \times 10^{-10} = \frac{x^2}{1.0 \times 10^{-4} - x} \approx \frac{x^2}{1.0 \times 10^{-4}}$$

$$x \approx 2.5 \times 10^{-7}\ M$$

Here $[H^+]$ from HCN is less than $10^{-6}$ $M$, so the full equation must be used:

$$6.2 \times 10^{-10} = K_a = \frac{[H^+]^2 - 10^{-14}}{1.0 \times 10^{-4} - \dfrac{[H^+]^2 - 10^{-14}}{[H^+]}}$$

Solve for $[H^+]$ by use of successive approximations. First determine a reasonable guess for $[H^+]$. Note from the above simple calculation that the $[H^+]$, ignoring the contribution from water, is $2.5 \times 10^{-7}$ $M$. Will the actual $[H^+]$ be larger or smaller than this?

It will be a little larger because of the contribution from $H_2O$. Guess $[H^+] = 3.0 \times 10^{-7}$ $M$. Substitute this value for $[H^+]$ into the denominator of the equation to give

$$K_a = 6.2 \times 10^{-10} = \frac{[H^+]^2 - 1.0 \times 10^{-14}}{1.0 \times 10^{-4} - \dfrac{(3.0 \times 10^{-7})^2 - 1.0 \times 10^{-14}}{3.0 \times 10^{-7}}}$$

$$6.2 \times 10^{-10} = \frac{[H^+]^2 - 1.0 \times 10^{-14}}{1.0 \times 10^{-4} - 2.67 \times 10^{-7}}$$

Now rearrange this equation so that a value for $[H^+]$ can be calculated:

$$[H^+]^2 = 6.2 \times 10^{-14} - 1.66 \times 10^{-16} + 1.0 \times 10^{-14} = 7.2 \times 10^{-14}$$

$$[H^+] = \sqrt{7.2 \times 10^{-14}} = 2.68 \times 10^{-7}$$

The original guessed value of $[H^+]$ was $3.0 \times 10^{-7}$. Since the calculated value and the guessed value do not agree, use $2.68 \times 10^{-7}$ as the new guessed value:

$$K_a = 6.2 \times 10^{-10} = \frac{[H^+]^2 - 1.0 \times 10^{-14}}{1.0 \times 10^{-4} - \dfrac{[H^+]^2 - 1.0 \times 10^{-14}}{[H^+]}}$$

$$= \frac{[H^+]^2 - 1.0 \times 10^{-14}}{1.0 \times 10^{-4} - \dfrac{(2.68 \times 10^{-7})^2 - 1.0 \times 10^{-14}}{2.68 \times 10^{-7}}}$$

Solving for $[H^+]$ gives

$$[H^+] = 2.68 \times 10^{-7} = 2.7 \times 10^{-7}\ M$$

Since the guessed value and the calculated value agree, this is the correct answer, which takes into account both the contribution $H^+$ from water and from HCN.

$$pH = -\log(2.7 \times 10^{-7}) = 6.57$$

**TEST 5.4** Calculate the $[H^+]$ in a $2.0 \times 10^{-4}\ M$ solution of phenol ($C_6H_5OH$, $K_a = 1.3 \times 10^{-10}$).

**A.** Do the problem the "normal" way.
**B.** How does the concentration of $H^+$ calculated in part A compare to $10^{-6}\ M$?
**C.** Must the full equation be used?
**D.** What might be a reasonable guess for the $[H^+]$?
**E.** Calculate the $[H^+]$.

## 5.4 Polyprotic Acids

### Introduction

Some acids have more than one acidic hydrogen. For example, the common strong acid, sulfuric acid, produces two protons per molecule:

$$H_2SO_4(aq) \rightleftarrows H^+(aq) + HSO_4^-(aq) \qquad K_{a_1} \text{ very large}$$
$$HSO_4^-(aq) \rightleftarrows H^+(aq) + SO_4^{2-}(aq) \qquad K_{a_2} = 1.2 \times 10^{-2}$$

Sulfuric acid is said to be a *diprotic acid*; the general term for acids that dissociate two or more $H^+$ ions per molecule is *polyprotic*. Two characteristics of polyprotic acids that normally can be assumed are

1. The protons dissociate stepwise, this is, two or more $H^+$ ions do not dissociate simultaneously.
2. The first dissociation occurs to the greatest extent, and subsequent dissociation steps have much smaller equilibrium constants.

To illustrate these concepts, consider phosphoric acid:

$$H_3PO_4(aq) \rightleftarrows H^+(aq) + H_2PO_4^-(aq) \qquad K_{a_1} = 7.5 \times 10^{-3}$$
$$H_2PO_4^-(aq) \rightleftarrows H^+(aq) + HPO_4^{2-}(aq) \qquad K_{a_2} = 6.2 \times 10^{-8}$$
$$HPO_4^{2-}(aq) \rightleftarrows H^+(aq) + PO_4^{3-}(aq) \qquad K_{a_3} = 4.8 \times 10^{-13}$$

Note that $H^+$ dissociation occurs in successive steps with decreasing equilibrium constants; the conjugate base of the first dissociation becomes the acid for the second dissociation, and so on.

Phosphoric acid illustrates a third characteristic that is usually (but not always) true for polyprotic acids: the successive $K_a$'s often differ by a factor of $10^3$ or more. For $H_3PO_4$

$$\frac{K_{a_1}}{K_{a_2}} = \frac{7.5 \times 10^{-3}}{6.2 \times 10^{-8}} = 1.2 \times 10^5$$

$$\frac{K_{a_1}}{K_{a_2}} = \frac{6.2 \times 10^{-8}}{4.8 \times 10^{-13}} = 1.3 \times 10^5$$

This characteristic causes pH calculations involving polyprotic acids to be much easier than might be initially expected. Under most circumstances, each successive acid (e.g., $H_3PO_4$, $H_2PO_4^-$, and $HPO_4^{2-}$) can be treated independently of the others. Thus the concepts learned for monoprotic acids can be applied to polyprotic acids with little modification.

## Calculations for Typical Polyprotic Acids

This section presents problems involving polyprotic acids with the three characteristics discussed above.

---

**Sample Exercise 5.6**

Calculate the pH of 1.00 *M* solution of $H_3PO_4$, for which $K_{a_1} = 7.5 \times 10^{-3}$, $K_{a_2} = 6.2 \times 10^{-8}$, and $K_{a_3} = 4.8 \times 10^{-13}$.

**Solution**

As always, first write the major species in solution:

$$H_3PO_4, H_2O$$

(None of the dissociation products of $H_3PO_4$ are written, because all of the $K_a$'s are small and these will be minor species.) Do this problem assuming that $H_3PO_4$ is the dominant acid ($K_{a_1} \gg K_{a_2} \gg K_{a_3}$). Under these conditions, this becomes a standard weak acid problem.

The dominant equilibrium is

$$H_3PO_4(aq) \rightleftarrows H^+(aq) + H_2PO_4^-(aq)$$

$$K_{a_1} = \frac{[H^+][H_2PO_4^-]}{[H_3PO_4]} = 7.5 \times 10^{-3}$$

| *Initial Concentrations (before any dissociation)* | | *Equilibrium Concentrations* |
|---|---|---|
| $[H_3PO_4]_0 = 1.00\ M$ | | $[H_3PO_4] = 1.00 - x$ |
| $[H_2PO_4^-]_0 = 0$ | Let $x$ mol/L $H_3PO_4$ dissociate $\longrightarrow$ | $[H_2PO_4^-] = x$ |
| $[H^+]_0 \approx 0$ (ignoring $H^+$ from $H_2O$) | | $[H^+] = x$ |

Solving for $x$ by substituting into $K_{a_1}$ and solving the easy way ($[H_3PO_4]_0 - x \approx [H_3PO_4]$) gives

$$x \approx 8.7 \times 10^{-2}$$

In this case,

$$\frac{x}{[H_3PO_4]_0} \times 100 = 8.7\%$$

so according to the 5% rule, the quadratic equation should be used. When this is done (do it for practice),

$$x = 8.3 \times 10^{-2}\ M = [H^+] = [H_2PO_4^-]$$

It has been assumed to this point that only the first dissociation of $H_3PO_4$ makes an important contribution to $[H^+]$. To check the validity of this assumption, substitute the calculated $[H^+]$ into the equilibrium expressions for the latter dissociations. For example, consider the second dissociation

$$H_2PO_4^-(aq) \rightleftarrows H^+(aq) + HPO_4^{2-}(aq)$$

where the equilibrium expression is

↙ $8.3 \times 10^{-2}\ M$ (from dissociation of $H_3PO_4$)

$$K_{a_2} = 6.2 \times 10^{-8} = \frac{[H^+][HPO_4^{2-}]}{[H_2PO_4^-]} = \frac{(8.3 \times 10^{-2})(HPO_4^{2-})}{(8.3 \times 10^{-2})}$$

↖ $8.3 \times 10^{-2}\ M$ (from dissociation of $H_3PO_4$)

$$[HPO_4^{2-}] = 6.2 \times 10^{-8} = K_{a_2}$$

Thus dissociation of $H_2PO_4^-$ to produce $H^+ + HPO_4^{2-}$ proceeds only to a very slight extent (only $6.2 \times 10^{-8}\ M\ HPO_4^{2-}$ is produced.)

A similar calculation can be performed involving the third dissociation:

↙ $8.3 \times 10^{-2}\ M$ (from first step)

$$4.8 \times 10^{-13} = K_{a_3} = \frac{[H^+][PO_4^{3-}]}{[H_2PO_4^{2-}]} = \frac{(8.3 \times 10^{-2})[PO_4^{3-}]}{6.2 \times 10^{-8}}$$

↖ $6.2 \times 10^{-8}\ M$ (from the second step)

$$[PO_4^{3-}] = \frac{(4.8 \times 10^{-13})(6.2 \times 10^{-8})}{8.3 \times 10^{-2}} = 3.6 \times 10^{-19}\ M$$

Thus the third dissociation, $HPO_4^{2-} \rightleftarrows H^+ + PO_4^{3-}$, occurs only to a minute extent.

All of this shows that, to calculate the pH of a 1.0 $M$ $H_3PO_4$ solution, only the first dissociation need be considered:

$$[H^+] = 8.3 \times 10^{-2}\ M$$

$$\text{pH} = 1.08$$

If the concentrations of $HPO_4^{2-}$ or $PO_4^{3-}$ are desired, they can be obtained from the $K_{a_2}$ or $K_{a_3}$ expressions as illustrated above.

---

**TEST 5.5** **A.** Calculate the pH of a 0.10 $M$ solution of carbonic acid ($H_2CO_3$), where

$$H_2CO_3(aq) \rightleftarrows H^+(aq) + HCO_3^-(aq) \qquad K_{a_1} = 4.3 \times 10^{-7}$$

$$HCO_3^-(aq) \rightleftarrows H^+(aq) + CO_3^{2-}(aq) \qquad K_{a_2} = 5.6 \times 10^{-11}$$

**B.** Calculate $[CO_3^{2-}]$ in this solution.

## Calculations Involving Sulfuric Acid

Sulfuric acid is a special case because the first acid ($H_2SO_4$) is a strong acid, but the second acid ($HSO_4^-$) is a weak acid. This special nature must be taken into account when equilibrium calculations are carried out.

---

**Sample Exercise 5.7**

Calculate the $[H^+]$ in a 1.00 $M$ $H_2SO_4$ solution, where

$$H_2SO_4(aq) \rightleftarrows H^+(aq) + HSO_4^-(aq) \qquad K_{a_1} \text{ very large}$$

$$HSO_4^-(aq) \rightleftarrows H^+(aq) + SO_4^{2-}(aq) \qquad K_{a_2} = 1.2 \times 10^{-2}$$

**Solution**

As mentioned above, the first acid is a strong acid and the second is weak. With this in mind, what are the major species in a 1.00 $M$ $H_2SO_4$ solution? Since $H_2SO_4$ dissociates essentially completely to $HSO_4^-$ and $H^+$, these species are

$$H^+, HSO_4^-, H_2O$$

The question that must now be answered is: To what extent does $HSO_4^-$ dissociate? To answer this question, do an equilibrium calculation where $HSO_4^-$ is the weak acid:

$$HSO_4^-(aq) \rightleftarrows H^+(aq) + SO_4^{2-}(aq)$$

$$1.2 \times 10^{-2} = \frac{[H^+][SO_4^{2-}]}{[HSO_4^-]}$$

*Initial Concentrations (before dissociation of $HSO_4^-$)*

$[HSO_4^-]_0 = 1.00\ M$ ← from complete dissociation of $H_2SO_4$

$[H^+]_0 = 1.00\ M$ ← from complete dissociation of $H_2SO_4$

$[SO_4^{2-}]_0 = 0$

Let $x$ mol/L $HSO_4^-$ dissociate →

*Equilibrium Concentrations*

$[HSO_4^-] = 1.00 - x$

$[H^+] = 1.00 + x$

$[SO_4^{2-}] = x$

$$1.2 \times 10^{-2} = K_{a_2} = \frac{[H^+][SO_4^{2-}]}{[HSO_4^-]} = \frac{(1.00 + x)(x)}{1.00 - x} \approx \frac{(1.00)(x)}{1.00}$$

$$x \approx 1.2 \times 10^{-2}\ M$$

Since $1.2 \times 10^{-2}$ is only 1.2% of 1.00, the simplifying approximation is valid.

$$[H^+] = 1.00 + x = 1.00 \times 0.012 = 1.01\ M$$

$$[HSO_4^-] = 1.00 - x = 1.00 - 0.012 = 0.99\ M$$

Note that in this case dissociation of $HSO_4^-$ contributes only 1% of the total $[H^+]$.

Even though the dissociation of $HSO_4^-$ makes only a small contribution to the total $[H^+]$, it is the only source of $SO_4^{2-}$ in the solution. Thus, to calculate the $[SO_4^{2-}]$, the equilibrium involving the dissociation of $HSO_4^-$ must be used:

$$HSO_4^-(aq) \rightleftarrows H^+(aq) + SO_4^{2-}(aq)$$

From the above calculations,

$$[H^+] = 1.01\ M$$

$$[HSO_4^-] = 0.99\ M$$

$$1.2 \times 10^{-2} = K_{a_2} = \frac{[H^+][SO_4^{2-}]}{[HSO_4^-]} = \frac{(1.01)[SO_4^{2-}]}{(0.99)}$$

$$[SO_4^{2-}] = \frac{0.99}{1.01}(1.2 \times 10^{-2}) = 1.17 \times 10^{-2}\ M = 1.2 \times 10^{-2}\ M$$

(rounding to the correct number of significant figures)

Sulfuric acid is the only common polyprotic acid for which the first acid is strong and the others are weak. The more common case is that in which each successive acid is a weak acid (e.g., $H_3PO_4$).

**TEST 5.6** Calculate the pH and the $[SO_4^{2-}]$ in a 5.0 $M$ $H_2SO_4$ solution. (For $H_2SO_4$, $K_{a_1}$ is very large and $K_{a_2} = 1.2 \times 10^{-2}$.)

## Calculations Involving Amphoteric Anions

So far, polyprotic acids have been dealt with using the procedures developed from the treatment of strong and weak monoprotic acids. However, there is one case commonly encountered in connection with polyprotic acids that does require some fresh thinking. This case occurs in solutions that contain, *as the only major species*, an "amphoteric anion." To understand this situation, consider $H_3PO_4$ again:

$$H_3PO_4(aq) \rightleftarrows H^+(aq) + H_2PO_4^-(aq)$$

$$H_2PO_4^-(aq) \rightleftarrows H^+(aq) + HPO_4^{2-}(aq)$$

$$HPO_4^{2-}(aq) \rightleftarrows H^+(aq) + PO_4^{3-}(aq)$$

Notice that

$H_3PO_4$ is an acid only

$PO_4^{3-}$ is a base only

but

$H_2PO_4^-$ is both an acid and a base (amphoteric)

$HPO_4^{2-}$ is both an acid and a base (amphoteric)

These latter two ions are examples of the amphoteric anions referred to above.

When a solution, such as a $NaH_2PO_4$ solution, contains $H_2PO_4^-$ as *the only major species*, both the acid and the base properties of $H_2PO_4^-$ must be taken into account for calculating the pH. This situation arises because $H_2PO_4^-$ can lose a proton:

$$H_2PO_4^-(aq) \rightleftarrows H^+(aq) + HPO_4^{2-}(aq)$$

or gain a proton:

$$H_2PO_4^-(aq) + H^+(aq) \rightleftarrows H_3PO_4(aq)$$

Of major species in solution, $H_2PO_4^-$ is both the best acid and the best base.

To treat this situation requires consideration of these equilibria simultaneously, which involves fairly complicated algebra that will not be considered here. Only the result will be given.

For every case to be encountered here, the pH of $NaH_2PO_4$ solution can be calculated from the expression

$$pH = \frac{pK_{a_1} + pK_{a_2}}{2}$$

where $pK_{a_1}$ and $pK_{a_2}$ are the negative logs of the dissociation constants for $H_3PO_4$ ($K_{a_1}$) and $H_2PO_4^-$ ($K_{a_2}$).

**Sample Exercise 5.8** Calculate the pH of a 1.0 *M* solution of $Na_2HPO_4$. (For $H_3PO_4$, $K_{a_1} = 7.5 \times 10^{-3}$, $K_{a_2} = 6.2 \times 10^{-8}$, $K_{a_3} = 4.8 \times 10^{-13}$.)

**Solution**

The major species in solution are

$$Na^+, HPO_4^{2-}, H_2O$$

This is an example of a solution containing the amphoteric anion $HPO_4^{2-}$. $HPO_4^{2-}$ is at the same time the best acid and the best base in the solution. Both properties must be considered to calculate the pH correctly. Use the formula involving the average of the p*K*'s. In this case, since $HPO_4^{2-}$ is the second amphoteric anion of $H_3PO_4$ ($H_2PO_4^-$ is the first), the $pK_{a_2}$ and $pK_{a_3}$ values must be used:

$$pH = \frac{pK_{a_2} + pK_{a_3}}{2} = \frac{7.21 + 12.32}{2} = 9.76$$

To summarize this situation:

| *Major Species* | *pH* |
|---|---|
| $H_2PO_4^-$ | $\frac{pK_{a_1} + pK_{a_2}}{2}$ |
| $HPO_4^{2-}$ | $\frac{pK_{a_2} + pK_{a_3}}{2}$ |

**Sample Exercise 5.9** Calculate the pH of a 0.10 *M* solution of $NaHCO_3$. (For $H_2CO_3$, $K_{a_1} = 4.3 \times 10^{-7}$ and $K_{a_2} = 5.6 \times 10^{-11}$.)

**Solution**

The major species in solution are $Na^+$, $HCO_3^-$, $H_2O$.

$HCO_3^-$ is the amphoteric anion associated with $H_2CO_3$. In a solution where $HCO_3^-$ is the major species,

$$pH = \frac{pK_{a_1} + pK_{a_2}}{2} = \frac{-\log(4.3 \times 10^{-7}) - \log(5.6 \times 10^{-11})}{2}$$

$$= \frac{6.37 + 10.25}{2} = \frac{16.62}{2} = 8.31$$

**Sample Exercise 5.10** Calculate the pH of a solution containing 0.100 *M* $NaHCO_3$ and 0.100 *M* $Na_2CO_3$. (For $H_2CO_3$, $K_{a_1} = 4.3 \times 10^{-7}$, $K_{a_2} = 5.6 \times 10^{-11}$.)

**Solution**

The major species in solution are: $Na^+$, $HCO_3^-$, $CO_3^{2-}$, $H_2O$.

Under these conditions, it might be tempting to assume, since $HCO_3^-$ is present, that the equation $pH = (pK_1 + pK_2)/2$ can be used. This formula *does not* give the correct answer in this case, because the solution contains large amounts of both $HCO_3^-$ and $CO_3^{2-}$. Therefore

$CO_3^{2-}$ is the strongest base

$HCO_3^-$ is the strongest acid

Thus only the acidic properties of $HCO_3^-$ need be considered in this case (since the much stronger base $CO_3^{2-}$ is present.) The equilibrium that will dominate in this solution is the one involving both $HCO_3^-$ and $CO_3^{2-}$:

$$HCO_3^-(aq) \rightleftarrows H^+(aq) + CO_3^{2-}(aq)$$

$$K_{a_2} = \frac{[H^+][CO_3^{2-}]}{[HCO_3^-]} = 5.6 \times 10^{-11}$$

| *Initial Concentrations* | | *Equilibrium Concentrations* |
|---|---|---|
| $[HCO_3^-]_0 = 0.100\ M$ | | $[HCO_3^-] = 0.100 - x$ |
| $[CO_3^{2-}]_0 = 0.100\ M$ | Let *x* mol/L $HCO_3^-$ dissociate $\longrightarrow$ | $[CO_3^{2-}] = 0.100 + x$ |
| $[H^+]_0 \approx 0$ (Ignoring $H^+$ from $H_2O$) | | $[H^+] = x$ |

$$5.6 \times 10^{-11} = K_{a_2} = \frac{[H^+][CO_3^{2-}]}{[HCO_3^-]} = \frac{(x)(0.100 + x)}{0.100 - x} \approx \frac{(x)(0.100)}{0.100}$$

$$x = 5.6 \times 10^{-11}\ M$$

$$pH = -\log(5.6 \times 10^{-11}) = 10.25$$

## Summary

When a solution contains an amphoteric anion as the *only major species,*

$$pH = \frac{pK_n + pK_m}{2}$$

where $n = 1$, $m = 2$ for the first amphoteric anion

$n = 2$, $m = 3$, for the second amphoteric anion (when the parent acid is triprotic)

When a solution contains large quantities of both an amphoteric anion and its conjugate acid or base, the pH is calculated using the appropriate dissociation equilibrium.

**TEST 5.7** Calculate the pH of each of the following solutions (for $H_3AsO_4$, $K_{a_1} = 5.0 \times 10^{-3}$, $K_{a_2} = 8.3 \times 10^{-8}$, $K_{a_3} = 6.0 \times 10^{-10}$):

**A.** 0.100 *M* $NaH_2AsO_4$
**B.** 0.100 *M* $Na_2HAsO_4$
**C.** 0.100 *M* in both $NaH_2AsO_4$ and $Na_2HAsO_4$

## Titration of Polyprotic Acids

Like the other pH calculations involving polyprotic acids, titrations can mostly be treated using the procedures for monoprotic acids.

**Sample Exercise 5.11**

Consider the titration of 100.00 mL of 1.00 *M* phthalic acid, $H_2C_8H_4O_4$ ($K_{a_1} = 1.3 \times 10^{-3}$, $K_{a_2} = 3.9 \times 10^{-6}$), with 1.00 *M* NaOH.

**A.** Calculate the pH of the solution before any NaOH is added. Do this as a test. (**Test 5.8**) (Hint: As always, first write down the major species and pick out the strongest acid.)

**B.** Calculate the pH after 25.0 mL (total) of 1.00 *M* NaOH has been added.

**Solution**

The mixed solution before any reaction occurs contains

$$Na^+, OH^-, H_2C_8H_4O_4, H_2O.$$

What reaction will occur? There is only one logical choice:

$$OH^-(aq) + H_2C_8H_4O_4(aq) \longrightarrow H_2O(l) + HC_8H_4O_4^-(aq)$$

Do the stoichiometry problem:

| | $OH^-$ | + $H_2C_8H_4O_4$ | $\longrightarrow$ $H_2O$ | + $HC_8H_4O_4^-$ |
|---|---|---|---|---|
| Before the reaction | 25.0 mmol | 100.0 mmol | | 0 |
| After the reaction | 0 | 75.0 mmol | | 25.0 mmol |

The stoichiometry problem is done. Now do the equilibrium problem.

In solution (after the above reaction has gone to completion) the species are

$$H_2C_8H_4O_4, \quad HC_8H_4O_4^-, Na^+, H_2O$$

$H_2C_8H_4O_4$ ↗ strongest acid; $HC_8H_4O_4^-$ ↖ strongest base

The equilibrium that will dominate in this solution is

$$H_2C_8H_4O_4(aq) \rightleftharpoons H^+(aq) + HC_8H_4O_4^-(aq)$$

Complete the problem to test your knowledge. (**Test 5.9**)

**C.** Calculate the pH after 50.0 mL (total) of 1.00 *M* NaOH has been added. (**Test 5.10**)

**D.** Calculate the pH after 100.0 mL of 1.00 *M* NaOH has been added.

**Solution**

This corresponds to the first equivalence point: 100.0 mmol of $OH^-$ have been added, which react with the original 100.0 mmol of $H_2C_8H_4O_4$ to produce 100.0 mmol of $HC_8H_4O_4^-$. Thus, after the titration reaction has run to completion, the solution contains

$$Na^+, HC_8H_4O_4^-, H_2O$$

How is this case handled?

It is tempting to note that $HC_8H_4O_4^-$ is a weak acid and to use $K_{a_2}$ to solve for the $[H^+]$. This will give an incorrect answer. What has been forgotten?

$HC_8H_4O_4^-$ is a base as well as an acid. (It is both the best acid and the best base in this solution.) Thus this is the special case described above (amphoteric anion):

$$\text{pH} = \frac{pK_{a_1} + pK_{a_2}}{2} = \frac{-\log(1.3 \times 10^{-3}) - \log(3.9 \times 10^{-6})}{2}$$

$$= \frac{2.89 + 5.41}{2} = \frac{8.30}{2} = 4.15$$

**E.** Calculate the pH after 125.0 mL of 1.00 *M* NaOH has been added.

**Solution**

This is 25.0 mL past the first equivalence point. Recall that

| $OH^-$ | + | $H_2C_8H_4O_4$ | $\longrightarrow$ | $H_2O$ | + | $HC_8H_4O_4^-$ |
|---|---|---|---|---|---|---|
| 100.0 mmol | | 100.0 mmol | | | | 100.0 mmol |
| ↖ requires 100.0 mL of 1.0 *M* NaOH | | | | | | |

When the $H_2C_8H_4O_4$ has been consumed, the $OH^-$ begins to react with the $HC_8H_4O_4^-$:

| | $OH^-$ | + | $HC_8H_4O_4^-$ | $\longrightarrow$ | $C_8H_4O_4^{2-}$ | + | $H_2O$ |
|---|---|---|---|---|---|---|---|
| Before the reaction | *25.0 mL × 1.0 *M* = 25.0 mmol | | 100.0 mmol | | 0 | | |
| After the reaction | 0 | | 75.0 mmol | | 25.0 mmol | | |

* The first 100.0 mL of NaOH was consumed by reaction with $H_2C_8H_4O_4$.

The stoichiometry problem is complete. Now do the equilibrium problem. The solution now contains

$$HC_8H_4O_4^-,\ C_8H_4O_4^{2-},\ Na^+,\ H_2O$$

What equilibrium will control the pH?

The strongest acid is $HC_8H_4O_4^-$, for which $C_8H_4O_4^{2-}$ is the conjugate base. Thus the equilibrium of interest is

$$HC_8H_4O_4^-(aq) \rightleftarrows H^+(aq) + C_8H_4O_4^{2-}(aq)$$

Now do the weak acid problem.

| *Initial Concentrations (before dissociation of $HC_8H_4O_4^-$)* | | *Equilibrium Concentrations* |
|---|---|---|
| $[HC_8H_4O_4^-]_0 = \dfrac{75.0}{100 + 125}$ | | $[HC_8H_4O_4^-] = \dfrac{75}{225} - x$ |
| $[C_8H_4O_4^{2-}]_0 = \dfrac{25.0}{100 + 125}$ | Let $x$ mol/L $HC_8H_4O_4^-$ dissociate $\longrightarrow$ | $[C_8H_4O_4^{2-}] = \dfrac{25}{225} + x$ |
| $[H^+]_0 \approx 0$ | | $[H^+] = x$ |

Substitute into $K_{a_2}$ and solve for $x$

$$x = 1.2 \times 10^{-5}\ M = [H^+];\ \text{pH} = 4.93$$

**F.** Calculate the pH after 150.0 mL of 1.00 *M* NaOH has been added. (**Test 5.11**)

**G.** Calculate the pH after 200.0 mL of 1.00 *M* has been added.

**Solution**

This is the second stoichiometric point; 100.0 mmol of $C_8H_4O_4^{2-}$ have been formed from the original 100.0 mmol of $H_2C_8H_4O_4$. In solution are $Na^+$, $C_8H_4O_4^{2-}$, $H_2O$. What equilibrium will dominate? Do the problem. (**Test 5.12**)

$\left(\text{Hint: } K_b \text{ for } C_8H_4O_4^{2-} = \dfrac{K_w}{K_{a_2}}\right)$

All of the concepts concerning titrations of polyprotic acids that have been discussed can be summarized on the following diagram for the titration of $H_3A$ with NaOH.

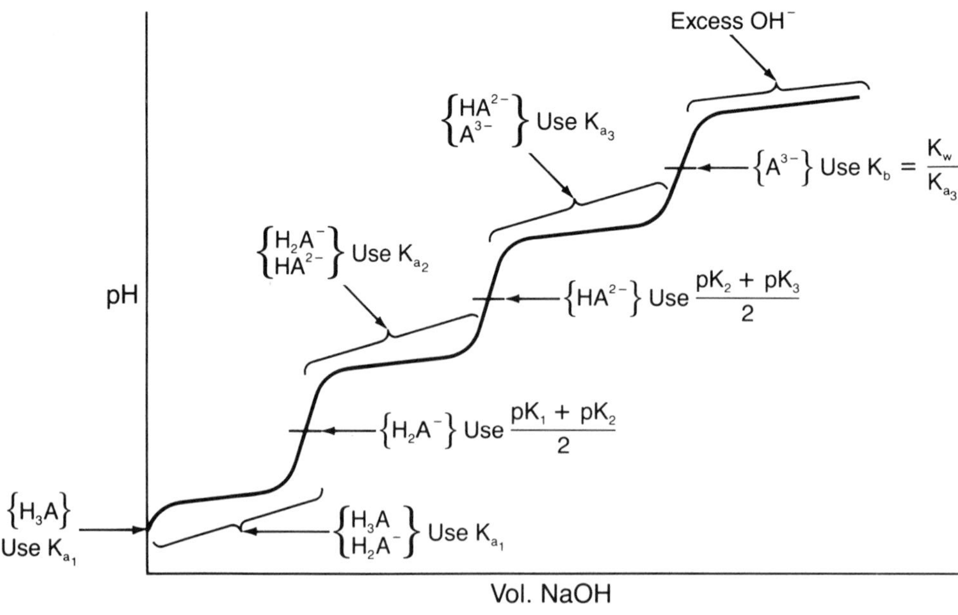


## 5.5 Equilibrium Calculations for Solutions Where Precipitation Occurs

In chapter 4, precipitation reactions occurring when appropriate solutions are mixed were considered. At that time, treatment of these solutions was limited to predicting whether or not precipitation would occur. In this section, equilibrium concentrations will be calculated.

**Sample Exercise 5.12** Calculate the equilibrium concentrations of $Pb^{2+}$ and $I^-$ in a solution prepared by mixing 100.0 mL of 0.0500 *M* $Pb(NO_3)_2$ and 200.0 mL of 0.100 *M* NaI. ($K_{sp}$ for $PbI_2$ is $1.4 \times 10^{-8}$.)

**Solution**

First determine whether $PbI_2(s)$ will form. To do this, calculate the original concentrations (before any reaction occurs) of $Pb^{2+}$ and $I^-$ in the mixed solution:

$$[Pb^{2+}]_{original} = \frac{\text{mmol } Pb^{2+}}{\text{mL of solution}} = \frac{(100.0 \text{ mL})(0.0500\ M)}{300.0 \text{ mL}} = 1.67 \times 10^{-2}\ M$$

$$[I^-]_{original} = \frac{\text{mmol } I^-}{\text{mL of solution}} = \frac{(200.0 \text{ mL})(0.100\ M)}{300.0 \text{ mL}} = 6.67 \times 10^{-2}\ M$$

For $PbI_2(s)$ ($PbI_2(s) \rightleftarrows Pb^{2+}(aq) + 2I^-(aq)$), the ion product expression is

$$Q = [Pb^{2+}]_0[I^-]_0^2 = (1.67 \times 10^{-2})(6.67 \times 10^{-2})^2$$
$$= 7.43 \times 10^{-5}$$

Thus $Q > K_{sp}$ and $PbI_2(s)$ will form.

Note that for $PbI_2(s)$ the equilibrium constant ($K_{sp}$) for the reaction

$$PbI_2(s) \rightleftarrows Pb^{2+}(aq) + 2I^-(aq)$$

is $1.4 \times 10^{-8}$. This means that the position of this equilibrium lies far to the left and that a solution cannot contain large concentrations of $Pb^{2+}$ and $I^-$. Another way of saying this is that $Pb^{2+}$ and $I^-$ will react essentially to completion.

As with any situation where a reaction goes essentially to completion, the math is easiest if the reaction is first considered to be complete and then the equilibrium calculations are made. So let the system go completely in the direction it tends toward, and then adjust it back to equilibrium.

The next step, then, is to let $Pb^{2+}$ and $I^-$ react to completion:

| | $Pb^{2+}$ | + | $2I^-$ | $\longrightarrow$ | $PbI_2(s)$ |
|---|---|---|---|---|---|
| Before the reaction | (100.0 mL)(0.0500 $M$) = 5.00 mmol | | (200.0 mL)(0.100 $M$) = 20.0 mmol | | amount of $PbI_2(s)$ does not influence the equilibrium |
| After the reaction | 0 | | 20.0 − 2(5.00) = 10.0 mmol | | |

At equilibrium the $[Pb^{2+}]$ is not really zero (the reaction does not quite go to completion). Correct for this by considering that some of the $PbI_2(s)$ formed in the reaction will dissolve:

$$PbI_2(s) \rightleftarrows Pb^{2+}(aq) + 2I^-(aq)$$

$$K_{sp} = [Pb^{2+}][I^-]^2 = 1.4 \times 10^{-8}$$

| *Initial Concentrations (after the reaction between $Pb^{2+}$ and $I^-$ has gone to completion)* | | *Equilibrium Concentrations* |
|---|---|---|
| $[Pb^{2+}]_0 = 0$ | | $[Pb^{2+}] = x$ |
| $[I^-]_0 = \dfrac{10.0 \text{ mmol}}{300.0 \text{ mL}} = 0.0333\ M$ | Let $x$ mol/L $PbI_2(s)$ dissolve to come to equilibrium $\longrightarrow$ | $[I^-] = 0.0333 + 2x$ |

$$1.4 \times 10^{-8} = K_{sp} = [Pb^{2+}][I^-]^2 = (x)(0.0333 + 2x)^2 \approx (x)(0.0333)^2$$

$x$ is expected to be very small ($PbI_2(s)$ is very insoluble). Thus

$$x \approx \frac{1.4 \times 10^{-8}}{(3.33 \times 10^{-2})^2} = 1.26 \times 10^{-5} \text{ mol/L}$$

Note that $0.033 \gg 2x = 2(1.26 \times 10^{-5})$. Thus at equilibrium

$$[Pb^{2+}] = x = 1.26 \times 10^{-5}\ M$$

$$[I^-] = 3.33 \times 10^{-2}\ M + 2x \approx 3.33 \times 10^{-2}\ M$$

In summary, the steps involved in this type of problem are

1. Determine whether or not precipitation occurs (use $Q$ to do this). If precipitation occurs:
2. Run the precipitation reaction to completion. (Do the stoichiometry problem.)
3. Compute the initial concentrations (after the precipitation reaction goes to completion).
4. Adjust the system to equilibrium.
   (a) Define $x$.
   (b) Represent the equilibrium concentrations in terms of $x$.
   (c) Solve for $x$ by plugging the equilibrium concentrations into the $K_{sp}$ expression (neglect $x$ where possible).

**Sample Exercise 5.13**

To help you appreciate the simplifications that are possible using the previous method, consider this problem using a different strategy: allow the system to come to equilibrium directly as it reacts to form $PbI_2(s)$.

| *Initial Concentrations (before any reaction occurs between $Pb^{2+}$ and $I^-$)* | | *Equilibrium Concentrations* |
|---|---|---|
| $[Pb^{2+}]_0 = 1.67 \times 10^{-2}\ M$ | Let $y$ mol/L $Pb^{2+}$ react with $2y$ mol/L of $I^-$ to come to equilibrium ⟶ | $[Pb^{2+}] = 1.67 \times 10^{-2} - y$ |
| $[I^-]_0 = 6.67 \times 10^{-2}\ M$ | | $[I^-] = 6.67 \times 10^{-2} - 2y$ |

$$1.4 \times 10^{-8} = K_{sp} = [Pb^{2+}][I^-]^2 = (1.67 \times 10^{-2} - y)(6.67 \times 10^{-2} - 2y)^2$$

Can $y$ be neglected? Clearly it cannot, because the $Pb^{2+}$ and the $I^-$ will react until the $Pb^{2+}$ is almost consumed ($PbI_2(s)$ being very insoluble); to solve the problem involves some very complicated algebra. Eventually one will obtain the same answer as by the previous method, but only after much harder math.

The point is that in dealing with substances that react almost to completion, it is best to run the reaction to completion first and then adjust to equilibrium. This is not the way nature does it, but the math is much easier and the answer will be correct.

**TEST 5.13** A solution is prepared by mixing 50.0 mL of $1.0 \times 10^{-3}\ M$ $Pb(NO_3)_2$ with 50.0 mL of $1.0 \times 10^{-2}\ M$ HCl. Calculate the concentrations of $Pb^{2+}$ and $Cl^-$ at equilibrium. (The $K_{sp}$ for $PbCl_2(s)$ is $1.0 \times 10^{-4}$.)

**A.** Does $PbCl_2(s)$ precipitate?
**B.** Calculate the concentrations of $Pb^{2+}$ and $Cl^-$.

**TEST 5.14** A solution is prepared by mixing 150.0 mL of $1.00 \times 10^{-2}\ M$ $Mg(NO_3)_2$ and 250.0 mL of $1.00 \times 10^{-1}\ M$ NaF. Calculate the concentrations of $Mg^{2+}$ and $F^-$ at equilibrium. (The $K_{sp}$ for $MgF_2(s)$ is $6.4 \times 10^{-9}$.)

**A.** Does $MgF_2(s)$ form?
**B.** Run the reaction to completion. (Do the stoichiometry problem.)
**C.** Calculate the initial concentrations of $Mg^{2+}$ and $F^-$ (after the precipitation reaction is run to completion but before some $MgF_2(s)$ redissolves to come to equilibrium).
**D.** Define $x$.
**E.** Represent the equilibrium concentrations in terms of $x$.
**F.** Solve for $x$ by plugging the equilibrium concentrations into the $K_{sp}$ expression.
**G.** Calculate the equilibrium concentrations of $Mg^{2+}$ and $F^-$.

## 5.6 Equilibria Involving Complex Ions

A complex ion is a charged species consisting of a metal ion surrounded by ligands. A ligand is typically an anion or neutral molecule that has an unshared electron pair (lone pair) that can be shared with an empty metal ion orbital to form a metal-ligand bond. Some common ligands are $H_2O$, $NH_3$, $Cl^-$, and $CN^-$.

The number of ligands attached to a metal ion is called the coordination number. The most common coordination numbers are six (as in $Co(OH_2)_6^{2+}$ and $Ni(NH_3)_6^{2+}$), four (as in $CoCl_4^{2-}$ and $Cu(NH_3)_4^{2+}$), and two (as in $Ag(NH_3)_2^+$), although others are known.

Metal ions add ligands one at a time in steps characterized by equilibrium constants called *formation constants*. For example, when solutions containing $Ag^+$ and $NH_3$ are mixed, the following reactions take place:

$$Ag^+(aq) + NH_3(aq) \rightleftarrows Ag(NH_3)^+(aq) \qquad K_1 = 2.1 \times 10^3$$

$$Ag(NH_3)^+(aq) + NH_3(aq) \rightleftarrows Ag(NH_3)_2^+(aq) \qquad K_2 = 8.2 \times 10^3$$

In a solution containing $Ag^+$ and $NH_3$, all of the species $NH_3$, $Ag^+$, $Ag(NH_3)^+$, and $Ag(NH_3)_2^+$ exist at equilibrium. Calculating the concentrations of all these components can be a complicated problem. However, in the most common case, where the total ligand concentration is much larger than the total metal ion concentration, certain approximations are possible that greatly simplify the algebra. This is the only case that will be considered here.

**Sample Exercise 5.14**

Calculate the equilibrium concentrations of $NH_3$, $Ag^+$, $Ag(NH_3)^+$, and $Ag(NH_3)_2^+$ in a solution prepared by mixing 100.0 mL of 2.0 $M$ $NH_3$ with 100.0 mL of $1.0 \times 10^{-3}$ $M$ $AgNO_3$.

**Solution**

Since equal volumes of the solutions have been mixed, the concentrations are halved. Thus, in the mixed solution before any reaction takes place,

$$[Ag^+]_{original} = \frac{1.0 \times 10^{-3}\ M}{2} = 5.0 \times 10^{-4}\ M$$

$$[NH_3]_{original} = \frac{2.0\ M}{2} = 1.0\ M$$

As usual, when two solutions are mixed, the first thing to do is to decide whether any reaction occurs that goes essentially to completion. If such a reaction exists, consider it completed first, and then make adjustments to allow the system to come to equilibrium, i.e., do the stoichiometry problem first and then worry about the equilibrium calculations. This strategy allows simplification of the algebra.

In this case, $Ag^+$ first reacts with $NH_3$ to form $Ag(NH_3)^+$. Since $K_1$ and the $[NH_3]_{original}$ are both large, this reaction would be expected to go essentially to completion.

The next step is the reaction of $Ag(NH_3)^+$ with $NH_3$ to form $Ag(NH_3)_2^+$. Again, since there is still a large amount of $NH_3$ available and since $K_2$ is large, this reaction will also essentially go to completion.

To summarize the situation:

1. $[NH_3]_{original} \gg [Ag^+]_{original}$
2. $K_1$ and $K_2$ are both large.

One concludes that at equilibrium most of the $Ag^+$ originally in the solution will be present as $Ag(NH_3)_2^+$. This is equivalent to writing the net reaction in the given solution as

$$Ag^+(aq) + 2NH_3(aq) \longrightarrow Ag(NH_3)_2^+(aq)$$

Under the conditions of this problem, $5 \times 10^{-4}$ mol/L $Ag^+$ reacts with $2(5 \times 10^{-4}$ mol/L) $NH_3$ to produce $5 \times 10^{-4}$ mol/L $Ag(NH_3)_2^+$. Thus the equilibrium concentration of $Ag(NH_3)_2$ can be estimated to be $5.0 \times 10^{-4}$ *M*.

What is the concentration of $NH_3$ at equilibrium? The solution originally contains 1.0 *M* $NH_3$, of which $2(5.0 \times 10^{-4}\ M)$ is consumed to form $5.0 \times 10^{-4}$ *M* $Ag(NH_3)_2^+$. At equilibrium

$$[NH_3] = 1.0\ M - 2(5.0 \times 10^{-4}\ M) \approx 1.0\ M$$

(Note that this discussion ignores the fact that $NH_3$ is a Brønsted base and reacts with $H_2O$. The amount of $NH_3$ consumed in this reaction is negligible.)

Next calculate the equilibrium concentration of $Ag(NH_3)^+$. To do this, recall the stepwise equilibria

$$Ag^+(aq) + NH_3(aq) \rightleftarrows Ag(NH_3)^+(aq) \qquad K_1 = 2.1 \times 10^3$$

$$Ag(NH_3)^+(aq) + NH_3(aq) \rightleftarrows Ag(NH_3)_2^+(aq) \qquad K_2 = 8.2 \times 10^3$$

Note that $Ag(NH_3)^+$ is involved in both equilibria. Which one should be used to calculate the $[Ag(NH_3)^+]$? The answer is that $K_2$ should be used, since the approximate equilibrium concentrations of $NH_3$ and $Ag(NH_3)_2^+$ are known.

$$8.2 \times 10^3 = K_2 = \frac{[Ag(NH_3)_2^+]}{[Ag(NH_3)^+][NH_3]} = \frac{(5.0 \times 10^{-4})}{[Ag(NH_3)^+](1.0)}$$

$$[Ag(NH_3)^+] = \frac{(5.0 \times 10^{-4})}{(8.2 \times 10^3)(1.0)} = 6.1 \times 10^{-8}\ M$$

Now calculate the equilibrium concentration of $Ag^+$, using $K_1$:

$$2.1 \times 10^3 = K_1 = \frac{[Ag(NH_3)^+]}{[Ag^+][NH_3]} = \frac{(6.1 \times 10^{-8})}{[Ag^+](1.0)}$$

$$[Ag^+] = \frac{(6.1 \times 10^{-8})}{(2.1 \times 10^3)(1.0)} = 2.9 \times 10^{-11}\ M$$

So far the assumption has been made that $Ag(NH_3)_2^+$ is the dominant silver-containing species in solution. The validity of this assumption can now be demonstrated. The calculated concentrations are

$$[Ag(NH_3)_2^+] = 5.0 \times 10^{-4}\ M$$

$$[Ag(NH_3)^+] = 6.1 \times 10^{-8}\ M$$

$$[Ag^+] = 2.9 \times 10^{-11}\ M$$

These values clearly support the conclusion that

$$[Ag(NH_3)_2^+] \gg [Ag(NH_3)^+] \gg [Ag^+]$$

Thus the assumption that $[Ag(NH_3)_2^+]$ is dominant is valid and the calculated concentrations are correct.

---

To review the strategy for dealing with a complex ion problem where $[L]_{original} \gg [M^{n+}]_{original}$ (L = ligand; $M^{n+}$ = metal ion):

1. Since the formation equilibrium constants are typically large, assume that the complex ion containing the most ligands will be dominant.
2. Calculate the equilibrium concentration of the ligand. This will be

$$[L]_{original} - \text{ligand consumed to form dominant species}$$

3. Calculate the concentrations of the intermediate species by using the stepwise formation constants.

---

**TEST 5.15** Calculate the concentrations of $Ag^+$, $Ag(S_2O_3)^-$ and $Ag(S_2O_3)_2^{3-}$ in a solution prepared by mixing 150.0 mL of $1.00 \times 10^{-3}$ $M$ $AgNO_3$ with 200.0 mL of 5.00 $M$ $Na_2S_2O_3$. The stepwise formation equilibria are

$$Ag^+(aq) + S_2O_3^{2-}(aq) \rightleftarrows Ag(S_2O_3)^-(aq) \qquad K_1 = 7.4 \times 10^8$$

$$Ag(S_2O_3)^-(aq) + S_2O_3^{2-}(aq) \rightleftarrows Ag(S_2O_3)_2^{3-}(aq) \qquad K_2 = 3.9 \times 10^4$$

**A.** Calculate the original concentrations of $Ag^+$ and $S_2O_3^{2-}$.
**B.** What will be the dominant silver-containing species? Calculate its approximate concentration at equilibrium.
**C.** What is the concentration of $S_2O_3^{2-}$ at equilibrium?
**D.** Calculate the equilibrium concentrations of $Ag(S_2O_3)^-$ and $Ag^+$.
**E.** Are all of the assumptions about relative concentrations valid?

---

**TEST 5.16** Calculate the equilibrium concentrations of $NH_3$, $Cu^{2+}$, $Cu(NH_3)^{2+}$, $Cu(NH_3)_2^{2+}$, $Cu(NH_3)_3^{2+}$, and $Cu(NH_3)_4^{2+}$ in a solution prepared by mixing 500.0 mL of 3.00 $M$ $NH_3$ with 500.0 mL of $2.00 \times 10^{-3}$ $M$ $Cu(NO_3)_2$. The stepwise equilibria are

$$Cu^{2+}(aq) + NH_3(aq) \rightleftarrows Cu(NH_3)^{2+}(aq) \qquad K_1 = 1.86 \times 10^4$$

$$Cu(NH_3)^{2+}(aq) + NH_3(aq) \rightleftarrows Cu(NH_3)_2^{2+}(aq) \qquad K_2 = 3.88 \times 10^3$$

$$Cu(NH_3)_2^{2+}(aq) + NH_3(aq) \rightleftarrows Cu(NH_3)_3^{2+}(aq) \qquad K_3 = 1.00 \times 10^3$$

$$Cu(NH_3)_3^{2+}(aq) + NH_3(aq) \rightleftarrows Cu(NH_3)_4^{2+}(aq) \qquad K_4 = 1.55 \times 10^2$$

---

## 5.7 Dissolving Solids That Are Very Insoluble in Water

### Introduction

On many occasions, a chemist must dissolve ionic compounds that are very insoluble in water. To understand how this might be done, consider the solubility equilibrium for a general salt, $MX(s)$, which is quite insoluble in water (very small $K_{sp}$ value):

$$MX(s) \rightleftarrows M^+(aq) + X^-(aq)$$

For significant quantities of $MX(s)$ to dissolve, this equilibrium must somehow be shifted to the right. This can be accomplished by lowering the concentration of either $M^+$ or $X^-$ by introducing a substance that will react with one of these ions.

If $X^-$ is a good base (that is, if HX is a weak acid), the usual strategy is to add a strong acid. This will produce the reaction

$$H^+(aq) + X^-(aq) \rightleftarrows HX(aq)$$

where the equilibrium position lies far to the right. As the $H^+$ reacts with $X^-$, lowering its concentration, the solubility equilibrium

$$MX(s) \rightleftarrows M^+(aq) + X^-(aq)$$

will be shifted to the right, thus increasing the solubility of MX(*s*). For example, the solubility of $Ag_3PO_4(s)$ is much greater in strong acid than in pure water. On the other hand, AgCl(*s*) has the same solubility in strong acid as in pure water. $PO_4^{3-}$ is a strong base and reacts with $H^+$ to form $HPO_4^{2-}$, whereas $Cl^-$ is a very weak base (HCl is a very strong acid in water) and virtually no HCl is formed.

How can one increase the solubility of a solid such as AgCl(*s*), which is very insoluble in both water and acid? The usual strategy in cases where $X^-$ is not a strong base is to add a ligand to the solution that forms a stable complex ion with the metal ion from the solid. For example, $Ag^+$ forms the stable complex ion $Ag(NH_3)_2^+$ in solutions containing $NH_3$. As a result, AgCl(*s*) is quite soluble in concentrated $NH_3$ solutions. The relevant reactions are

$$AgCl(s) \rightleftarrows Ag^+(aq) + Cl^-(aq)$$

$$Ag^+(aq) + NH_3(aq) \rightleftarrows Ag(NH_3)^+(aq)$$

$$Ag(NH_3)^+(aq) + NH_3(aq) \rightleftarrows Ag(NH_3)_2^+(aq)$$

The $Ag^+$ produced by dissolving AgCl(*s*) is combined with the $NH_3$ to form $Ag(NH_3)_2^+$, which causes more AgCl(*s*) to dissolve, until

$$[Ag^+][Cl^-] = K_{sp} = 1.6 \times 10^{-10}$$

↖ free $Ag^+$ only; $Ag(NH_3)^+$ and $Ag(NH_3)_2^+$ not included

## Strategies for Dissolving Water-Insoluble Ionic Solids

Consider the salt MX(*s*).

1. If $X^-$ is a strong base (HX is a weak acid), add $H^+$ to form HX, thus lowering the $X^-$ concentration. Examples of common anions that are good bases are $OH^-$, $S^{2-}$, $CO_3^{2-}$, $PO_4^{3-}$, $C_2O_4^{2-}$, and $CrO_4^{2-}$. Salts containing these anions are much more soluble in acid than in pure water.
2. If $X^-$ is not a strong base (HX is a strong acid), add a ligand that will form stable complex ions with $M^+$.

## Equilibrium Calculations

These cases will now be treated quantitatively.

*Dissolving Salts in Strong Acid Solution*

**Sample Exercise 5.15** Calculate the solubility of $Ag_3PO_4(s)$ ($K_{sp} = 1.8 \times 10^{-18}$) in a 10.0 *M* $HNO_3$ solution. (For $H_3PO_4$, $K_{a_1} = 7.5 \times 10^{-3}$, $K_{a_2} = 6.2 \times 10^{-8}$, and $K_{a_3} = 4.8 \times 10^{-13}$.)

**Solution**

This is a complicated problem, and must be thought about carefully. First, review the calculation of the solubility of $Ag_3PO_4(s)$ in pure water. The equilibrium is

$$Ag_3PO_4(s) \rightleftarrows 3Ag^+(aq) + PO_4^{3-}(aq)$$

$$K_{sp} = [Ag^+]^3[PO_4^{3-}] = 1.8 \times 10^{-18}$$

Let $x$ = solubility in water. Then

$$[Ag^+] = 3x$$

$$[PO_4^{3-}] = x$$

$$1.8 \times 10^{-18} = K_{sp} = (3x)^3(x) = 27x^4$$

$$x^4 = \frac{1.8 \times 10^{-18}}{27} = 6.7 \times 10^{-20}$$

$$x = 1.6 \times 10^{-5}\ \text{mol/L} = \text{solubility in pure water}$$

Those of you who are thinking carefully about this problem may wonder how the strong basicity of $PO_4^{3-}$ (the $K_a$ for $HPO_4^{2-}$ is only $4.8 \times 10^{-13}$) may affect the solubility of $Ag_3PO_4(s)$ in water. (In the problem above, the properties of $PO_4^{3-}$ as a base have been ignored.) Actually, the solubility of $Ag_3PO_4(s)$ in water is greater than the value calculated above, because some of the $PO_4^{3-}$ from the dissolved $Ag_3PO_4$ reacts with water. The equilibria involved are

$$Ag_3PO_4(s) \rightleftarrows 3Ag^+(aq) + PO_4^{3-}(aq)$$

$$PO_4^{3-}(aq) + H_2O(l) \rightleftarrows HPO_4^{2-}(aq) + OH^-(aq)$$

The actual solubility of $Ag_3PO_4(s)$ in water is greater than $1.6 \times 10^{-5}$ mol/L, because the second reaction lowers the $[PO_4^{3-}]$ and thus shifts the first equilibrium to the right.

If you are interested in the details of this problem, it is considered in the following optional test.

**TEST 5.17 (optional)** Calculate accurately the solubility of $Ag_3PO_4(s)$ ($K_{sp} = 1.8 \times 10^{-18}$) in pure water. (For $H_3PO_4$, $K_{a_1} = 7.5 \times 10^{-3}$, $K_{a_2} = 6.2 \times 10^{-8}$, and $K_{a_3} = 4.8 \times 10^{-13}$.)

Now calculate the solubility of $Ag_3PO_4(s)$ in 10.0 $M$ $HNO_3$. The reactions that must be considered are

$$Ag_3PO_4(s) \rightleftarrows 3Ag^+(aq) + PO_4^{3-}(aq) \qquad K_{sp} = 1.8 \times 10^{-18}$$

$$H^+(aq) + PO_4^{3-}(aq) \rightleftarrows HPO_4^{2-}(aq) \qquad K = \frac{1}{K_{a_3}} = 2.1 \times 10^{12}$$

$$HPO_4^{2-}(aq) + H^+(aq) \rightleftarrows H_2PO_4^-(aq) \qquad K = \frac{1}{K_{a_2}} = 1.6 \times 10^{7}$$

$$H_2PO_4^-(aq) + H^+(aq) \rightleftarrows H_3PO_4(aq) \qquad K = \frac{1}{K_{a_1}} = 1.3 \times 10^{2}$$

The $Ag_3PO_4(s)$ dissolves to produce $PO_4^{3-}$, which reacts with the $H^+$ from the 10.0 $M$ $HNO_3$.

Note that the equilibrium of each of the protonation steps lies far to the right. Thus, in the presence of a high concentration of $H^+$, it is reasonable to assume that almost all of the $PO_4^{3-}$ from the $Ag_3PO_4$ ends up as $H_3PO_4$. To simplify this problem, assume that $H_3PO_4$ is the dominant phosphate-containing species in the solution. This is the same as assuming that the net reaction is

$$Ag_3PO_4(s) + 3H^+(aq) \longrightarrow H_3PO_4(aq) + 3Ag^+(aq)$$

which is the sum of the four reactions listed above. The $K$ for this equilibrium is the product of the $K$ values for the four reactions:

$$K = \frac{[H_3PO_4][Ag^+]^3}{[H^+]^3} = \frac{K_{sp}}{K_{a_1} \cdot K_{a_2} \cdot K_{a_3}} = 8.1 \times 10^{3}$$

Let $x$ = solubility of $Ag_3PO_4$. Then $xAg_3PO_4$ reacts with $3xH^+$ to produce $xH_3PO_4$ plus $3xAg^+$. At equilibrium (assuming 1 L of solution),

$$[H_3PO_4] = x$$

$$[Ag^+] = 3x$$

$$[H^+] = 10.0 - \text{amount consumed} = 10.0 - 3x$$

$$K = 8.1 \times 10^3 = \frac{(x)(3x)^3}{(10.0 - 3x)^3}$$

At this point a crucial question arises: Is $10 \gg 3x$? The answer is "probably not." Note that the $K$ for this reaction is quite large. A large amount of $H^+$, therefore, will be consumed as the reaction comes to equilibrium. The equation can be solved by trial and error, or by adopting a different strategy. Since the reaction

$$Ag_3PO_4(s) + 3H^+(aq) \longrightarrow H_3PO_4(aq) + 3Ag^+(aq)$$

goes nearly to completion, a more reasonable strategy is to assume the reaction goes to completion and then adjust it to equilibrium.

| | $Ag_3PO_4$ + | $3H^+$ | $\longrightarrow$ | $H_3PO_4$ + | $3Ag^+$ |
|---|---|---|---|---|---|
| Before the reaction | excess | 10.0 mol | | 0 | 0 |
| After the reaction | excess | 0 | | 3.33 mol | 10.0 mol |

Now let the system come back to equilibrium by considering the reverse reaction:

| *Initial Concentrations (above reaction has gone to completion)* | | *Equilibrium Concentrations* |
|---|---|---|
| $[Ag^+]_0 = 10.0\ M$ | Let $y$ mol/L $H_3PO_4$ $\longrightarrow$ react with $Ag^+$ | $[Ag^+] = 10.0 - 3y$ |
| $[H_3PO_4]_0 = 3.33\ M$ | | $[H_3PO_4] + 3.33 - y$ |
| $[H^+]_0 = 0$ | | $[H^+] = 3y$ |

Now substitute into the equilibrium expression:

$$K = 8.1 \times 10^3 = \frac{[Ag^+]^3[H_3PO_4]}{[H^+]^3} = \frac{(10 - 3y)^3(3.33 - y)}{(3y)^3}$$

Assume $10 \gg 3y$ or $3.33 \gg y$ so that

$$8.1 \times 10^3 \approx \frac{(10)^3(3.33)}{(3y)^3} = \frac{3.33 \times 10^3}{27y^3}$$

$$y^3 = \frac{3.33 \times 10^3}{(27)(8.1 \times 10^3)}$$

$$y = 0.25\ \text{mol/L}$$

Note that $3y = 0.75$ mol/L. Comparing $3y$ to 10.0:

$$\frac{0.75}{10.0} \times 100 = 7.5\%$$

To continue to abide by the "5% rule," this should be solved more exactly. Solving by trial and error produces a value of $y = 0.23$ mol/L.

Now check the assumption that $H_3PO_4$ is the dominant species.

Using $K_{a_1}$:

$$K_{a_1} = 7.5 \times 10^{-3} = \frac{[H^+][H_2PO_4^-]}{[H_3PO_4]}$$

$$[H^+] = 3y = 0.69\ M$$

$$\frac{[H_3PO_4]}{[H_2PO_4^-]} = \frac{[H^+]}{K_{a_1}} = \frac{0.69}{7.5 \times 10^{-3}} = \frac{92}{1}$$

Thus $[H_3PO_4] \gg [H_2PO_4^-]$.

Using $K_{a_2}$:

$$K_{a_2} = \frac{[H^+][HPO_4^{2-}]}{[H_2PO_4^-]} = 6.2 \times 10^{-8}$$

$$\frac{[H_2PO_4^-]}{[HPO_4^{2-}]} = \frac{[H^+]}{K_{a_2}} = \frac{0.69}{6.2 \times 10^{-8}} = 1.1 \times 10^7$$

Thus $[H_2PO_4^-] \gg [HPO_4^{2-}]$.

Using $K_{a3}$:

$$K_{a_3} = \frac{[H^+][PO_4^{3-}]}{[HPO_4^{2-}]} = 4.8 \times 10^{-13}$$

$$\frac{[HPO_4^{2-}]}{[PO_4^{3-}]} = \frac{[H^+]}{K_{a_3}} = \frac{0.69}{4.8 \times 10^{-13}} = 1.4 \times 10^{12}$$

Thus $[HPO_4^{2-}] \gg [PO_4^{3-}]$.

To summarize: $[H_3PO_4] \gg [H_2PO_4^-] \gg [HPO_4^{2-}] \gg [PO_4^{3-}]$.

The original assumption that $H_3PO_4$ is dominant is valid; that is, the net reaction is

$$Ag_3PO_4(s) + 3H^+(aq) \longrightarrow H_3PO_4(aq) + 3Ag^+(aq)$$

The total solubility is equal to the concentration of $H_3PO_4$ at equilibrium. From the above calculations,

$$[H_3PO_4] = 3.33\ M - y = 3.33 - 0.23 = 3.10\ \text{mol/L}$$

Therefore the solubility of $Ag_3PO_4(s)$ in 10.0 $M$ $HNO_3$ is 3.10 mol/L. Compare this with a solubility of $\sim 10^{-5}$ mol/L in pure water. The acid dramatically increases the solubility of $Ag_3PO_4(s)$.

*Strategies for Attacking This Type of Problem*

1. Write the reactions that occur.
2. Determine whether one species is dominant.
3. If so, write the net reaction that occurs.
4. Solve the equilibrium problem using the net reaction. (When equations are added, $K$ for the net reaction is the product of the individual $K$ values.)

**TEST 5.18** Calculate the solubility of $CuS(s)$ ($K_{sp} = 8.5 \times 10^{-45}$) in 10.0 $M$ strong acid. (For $H_2S$, $K_{a_1} = 1.02 \times 10^{-7}$, $K_{a_2} = 1.29 \times 10^{-13}$.)

**A.** Write the steps (reactions) that occur as $CuS(s)$ dissolves in this solution.

**B.** Determine which species containing S is most likely to be dominant.

**C.** Write the net reaction that most likely occurs.
**D.** Write the equilibrium expression for the net reaction and compute the $K$ value.
**E.** Do the equilibrium calculations. (Hint: Let $x$ = solubility of $CuS(s)$.)

*Dissolving Salts in Solutions Containing a Ligand*

**Sample Exercise 5.16**

Calculate the solubility of $AgCl(s)$ ($K_{sp} = 1.6 \times 10^{-10}$) in 10.0 $M$ $NH_3$. $Ag^+$ reacts with $NH_3$ to form $Ag(NH_3)^+$ ($K_1 = 2.1 \times 10^3$) and $Ag(NH_3)_2^+$ ($K_2 = 8.2 \times 10^3$).

**Solution**

As the $AgCl(s)$ dissolves, the $Ag^+$ reacts with $NH_3$ to produce $Ag(NH_3)^+$ and $Ag(NH_3)_2^+$. The relevant reactions are

$$AgCl(s) \rightleftarrows Ag^+(aq) + Cl^-(aq) \qquad K_{sp} = 1.6 \times 10^{-10}$$

$$Ag^+(aq) + NH_3(aq) \rightleftarrows Ag(NH_3)^+(aq) \qquad K_1 = 2.1 \times 10^3$$

$$Ag(NH_3)^+(aq) + NH_3(aq) \rightleftarrows Ag(NH_3)_2^+(aq) \qquad K_2 = 8.2 \times 10^3$$

Note that the equilibrium in each of the complexation reactions lies far to the right. Thus, in the presence of excess $NH_3$, $Ag(NH_3)_2^+$ would be expected to be the dominant silver-containing species. This means that the net reaction will be

$$AgCl(s) + 2NH_3(aq) \rightleftarrows Ag(NH_3)_2^+(aq) + Cl^-(aq)$$

which is the sum of the three reactions given above.

$$K = \frac{[Ag(NH_3)_2^+][Cl^-]}{[NH_3]^2} = K_{sp}K_1K_2$$

$$= (1.6 \times 10^{-10})(2.1 \times 10^3)(8.2 \times 10^3)$$

$$= 2.8 \times 10^{-3}$$

Let $x$ = solubility of AgCl. Then $xAgCl(s)$ reacts with $2xNH_3$ to produce $xAg(NH_3)_2^+$ plus $xCl^-$. Thus, at equilibrium,

$$[Cl^-] = x$$

$$[Ag(NH_3)_2^+] = x$$

$$[NH_3] = \text{initial amount} - \text{amount consumed} = 10.0 - 2x$$

$$K = 2.8 \times 10^{-3} = \frac{[Ag(NH_3)_2^+][Cl^-]}{[NH_3]^2} = \frac{(x)(x)}{(10.0 - 2x)^2}$$

Since $K$ is small, assume that $10.0 \gg 2x$, and thus that $10.0 - 2x \approx 10.0$:

$$2.8 \times 10^{-3} = \frac{(x)(x)}{(10.0 - 2x)^2} \approx \frac{x^2}{(10.0)^2}$$

$$x^2 \approx (10.0)^2(2.8 \times 10^{-3}) = 2.8 \times 10^{-1}$$

$$x \approx 0.53 \text{ mol/L}$$

Check the assumption: $2x = 2(0.53) = 1.06$:

$$\frac{2x}{10} \times 100 = \frac{1.06}{10} \times 100 = 10.6\%$$

Thus the assumption that $10.0 - 2x \approx 10.0$ is *not* valid. The equation must be solved exactly (take the square root of both sides). This gives $x = 0.48$ mol/L.

$$[NH_3] = 10.0 - 2x = 10.0 - 0.96 = 9.0\ M$$

Now check the assumption that $Ag(NH_3)_2^+$ is the dominant species.
Using $K_2$:

$$K_2 = \frac{[Ag(NH_3)_2^+]}{[Ag(NH_3)^+][NH_3]}$$

$$\frac{[Ag(NH_3)_2^+]}{[Ag(NH_3)^+]} = [NH_3] \cdot K_2 = (9.0)(8.2 \times 10^3) = 7.4 \times 10^4$$

Thus $[Ag(NH_3)_2^+] \gg [Ag(NH_3)^+]$.
Using $K_1$:

$$K_1 = \frac{[Ag(NH_3)^+]}{[Ag^+][NH_3]}$$

$$\frac{[Ag(NH_3)^+]}{[Ag^+]} = K_1[NH_3] = (2.1 \times 10^3)(9.0) = 1.9 \times 10^4$$

Thus $[Ag(NH_3)^+] \gg [Ag^+]$.
Summary: $[Ag(NH_3)_2^+ \gg [Ag(NH_3)^+] \gg [Ag^+]$. $Ag(NH_3)_2^+$ is the dominant species. The solubility of AgCl($s$) in 10.0 $M$ $NH_3$ is 0.48 mol/L.

**TEST 5.19** Calculate the solubility of AgI($s$) ($K_{sp} = 1.5 \times 10^{-16}$) in a 5.0 $M$ $Na_2S_2O_3$ solution. $Ag^+$ reacts with $S_2O_3^{2-}$ to form $Ag(S_2O_3)^-$ ($K_1 = 7.4 \times 10^8$) and $Ag(S_2O_3)_2^{3-}$ ($K_2 = 3.9 \times 10^4$).

**A.** Write the series of reactions that occurs when AgI($s$) dissolves in this solution.
**B.** Decide which silver-containing species will be dominant.
**C.** Write the net reaction that occurs in this solution.
**D.** Write the equilibrium expression for the net reaction and calculate the value of $K$.
**E.** Solve the equilibrium problem.

## EXERCISES

1. Calculate the pH of a 0.0100 $M$ solution of chloroacetic acid, $HC_2H_2O_2Cl$ ($K_a = 1.4 \times 10^{-3}$).
2. Calculate the pH of a $5.0 \times 10^{-8}$ $M$ $HNO_3$ solution.
3. Calculate the pH of a $1.0 \times 10^{-2}$ $M$ solution of $Fe(NO_3)_3$. ($Fe^{3+}$ is hydrated and the $K_a$ for $Fe(OH_2)_6^{3+}$ is $6.0 \times 10^{-3}$.)
4. Calculate the pH of a $5.0 \times 10^{-5}$ $M$ solution of HCN ($K_a = 6.2 \times 10^{-10}$).
5. Calculate the pH of a 1.00 $M$ solution of oxalic acid, $H_2C_2O_4$ ($K_{a_1} = 6.5 \times 10^{-2}$, $K_{a_2} = 6.1 \times 10^{-5}$).
6. Calculate the pH for each of the following:
   (a) 0.100 $M$ $H_3AsO_4$ ($K_{a_1} = 5.0 \times 10^{-3}$, $K_{a_2} = 8.3 \times 10^{-8}$, $K_{a_3} = 6.0 \times 10^{-10}$).
   (b) 0.100 $M$ $NaH_2AsO_4$.
   (c) A solution prepared by mixing 500.0 mL of 0.100 $M$ NaOH and 500.0 mL of 0.200 $M$ $H_3AsO_4$.
   (d) 0.100 $M$ $Na_3AsO_4$.

**7.** Calculate the pH of a 0.90 *M* $H_2SO_4$ solution ($K_{a_2} = 1.2 \times 10^{-2}$).

**8.** Consider 75.0 mL of a 0.30 *M* $H_3PO_4$ solution (for $H_3PO_4$, $K_{a_1} = 7.5 \times 10^{-3}$, $K_{a_2} = 6.2 \times 10^{-8}$, $K_{a_3} = 4.8 \times 10^{-13}$).
(a) Calculate the pH of this solution.
(b) Calculate the pH after 10.0 mL of 1.0 *M* NaOH is added.

**9.** Calculate the pH of a solution containing $1.00 \times 10^{-3}$ *M* HCl and 1.0 *M* acetic acid ($K_a = 1.8 \times 10^{-5}$).

**10.** A solution is prepared by mixing 100.0 mL of 0.100 *M* $KIO_3$ and 100.0 mL of $1.00 \times 10^{-2}$ *M* $Ce(NO_3)_3$. Calculate the concentrations of $Ce^{3+}$ and $IO_3^-$ at equilibrium. (The $K_{sp}$ for $Ce(IO_3)_3(s)$ is $3.5 \times 10^{-10}$.)

**11.** A solution is prepared by mixing 100.0 mL of $1.00 \times 10^{-4}$ *M* $Be(NO_3)_2$ and 100.0 mL of 8.00 *M* NaF.

$$Be^{2+}(aq) + F^-(aq) \rightleftharpoons BeF^+(aq) \qquad K_1 = 7.9 \times 10^4$$

$$BeF^+(aq) + F^-(aq) \rightleftharpoons BeF_2(aq) \qquad K_2 = 5.8 \times 10^3$$

$$BeF_2(aq) + F^-(aq) \rightleftharpoons BF_3^-(aq) \qquad K_3 = 6.1 \times 10^2$$

$$BeF_3^-(aq) + F^-(aq) \rightleftharpoons BeF_4^{2-}(aq) \qquad K_4 = 2.7 \times 10^1$$

Calculate the equilibrium concentrations of $F^-$, $Be^{2+}$, $BeF^+$, $BeF_2$, $BeF_3^-$, and $BeF_4^{2-}$ in this solution.

**12.** Calculate the solubility of AgCN(*s*) ($K_{sp} = 2.2 \times 10^{-12}$) in a solution containing 1.0 *M* $H^+$. (The $K_a$ for HCN is $6.2 \times 10^{-10}$.)

**13.** Calculate the solubility of HgS(*s*) ($K_{sp} = 1.0 \times 10^{-52}$) in a solution containing 10.0 *M* $H^+$. (For $H_2S$, $K_{a_1} = 1.02 \times 10^{-7}$, $K_{a_2} = 1.29 \times 10^{-13}$.)

**14.** Calculate the solubility of $Pb_3(PO_4)_2(s)$ ($K_{sp} = 1.0 \times 10^{-42}$) in 1.0 *M* $HNO_3$. (For $H_3PO_4$, $K_{a_1} = 7.5 \times 10^{-3}$, $K_{a_2} = 6.2 \times 10^{-8}$, $K_{a_3} = 4.8 \times 10^{-13}$.)

**15.** Calculate the solubility of AgI(*s*) ($K_{sp} = 1.5 \times 10^{-16}$) in 5.0 *M* $NH_3$. $Ag^+$ reacts with $NH_3$ to form $Ag(NH_3)^+$ ($K_1 = 2.1 \times 10^3$) and $Ag(NH_3)_2^+$ ($K_2 = 8.2 \times 10^3$).

**16.** Calculate the solubility of ZnS(*s*) ($K_{sp} = 1.2 \times 10^{-23}$) in 5.0 *M* $NH_3$. $Zn^{2+}$ reacts with $NH_3$ to form $Zn(NH_3)^{2+}$ ($K_1 = 1.5 \times 10^2$), $Zn(NH_3)_2^{2+}$ ($K_2 = 1.8 \times 10^2$), $Zn(NH_3)_3^{2+}$ ($K_3 = 2.0 \times 10^2$), $Zn(NH_3)_4^{2+}$ ($K_4 = 1.0 \times 10^2$). Ignore all Brønsted acid-base reactions.

**Multiple Choice Questions**

**17.** What is the $[H^+]$ in a $1.00 \times 10^{-3}$ *M* solution of phenol ($C_6H_5OH$, $K_a = 1.3 \times 10^{-10}$)?
(a) $1.0 \times 10^{-7}$ *M*
(b) $4.7 \times 10^{-7}$ *M*
(c) $1.0 \times 10^{-3}$ *M*
(d) $3.7 \times 10^{-7}$ *M*
(e) None of these

**18.** The pH in a solution of 1.0 *M* $H_2X$ ($K_1 = 1.0 \times 10^{-6}$, $K_2 = 1.0 \times 10^{-10}$) is
(a) 5.00
(b) 7.00
(c) 6.00
(d) 3.00
(e) None of these

**19.** Calculate the solubility of MnS in 0.20 *M* $H^+$. For MnS(*s*), $K_{sp} = 1.4 \times 10^{-15}$, and for $H_2S$, $K_1 = 1.02 \times 10^{-7}$, $K_2 = 1.29 \times 10^{-13}$
(a) 0.10 mol/L
(b) 0.20 mol/L
(c) $2.8 \times 10^{-7}$ mol/L
(d) $1.4 \times 10^4$ mol/L
(e) None of these

**20.** The equilibrium $[H^+]$ in question 19 is
(a) 0
(b) 0.10 $M$
(c) $1.0 \times 10^{-7}$ $M$
(d) $1.8 \times 10^{-4}$ $M$
(e) $3.0 \times 10^{-4}$ $M$

For exercises 21 and 22, consider a solution made up by mixing 50.0 mL of $2.0 \times 10^{-4}$ $M$ $CuNO_3$ and 50.0 mL of 4.0 $M$ NaCN. $Cu^+$ reacts with $CN^-$ to form the complex ion $Cu(CN)_3^{2-}$ where

$$Cu^+(aq) + CN^-(aq) \rightleftarrows CuCN(aq) \qquad K_2 = 1.0 \times 10^2$$

$$CuCN(aq) + CN^-(aq) \rightleftarrows Cu(CN)_2^-(aq) \qquad K_2 = 1.0 \times 10^3$$

$$Cu(CN)_2^-(aq) + CN^-(aq) \rightleftarrows Cu(CN)_3^{2-}(aq) \qquad K_3 = 1.0 \times 10^4$$

**21.** The concentration of $Cu^+$ at equilibrium is
(a) $2.0 \times 10^{-4}$ $M$
(b) $1.0 \times 10^{-4}$ $M$
(c) $1.3 \times 10^{-14}$ $M$
(d) $5.0 \times 10^{-14}$ $M$
(e) None of these

**22.** Calculate the solubility of CuBr($s$) ($K_{sp} = 1.0 \times 10^{-5}$) in 1.0 L of 1.0 $M$ NaCN.
(a) 1.0 mol
(b) $1.0 \times 10^{-6}$ mol
(c) 0.33 mol
(d) $1.0 \times 10^3$ mol
(e) None of these

**23.** 50.00 mL of a 1.00 $M$ solution of a diprotic acid, $H_2A$ ($K_1 = 1.0 \times 10^{-6}$ and $K_2 = 1.0 \times 10^{-10}$), is being titrated with 2.00 $M$ NaOH. How many mL of 2.00 $M$ NaOH must be added to reach a pH of 10.00?
(a) 0
(b) 12.5 mL
(c) 25.0 mL
(d) 37.5 mL
(e) None of these

**24.** The $[H^+]$ in a solution containing $1.0 \times 10^{-6}$ $M$ HOCl ($K_a = 4.0 \times 10^{-8}$) is
(a) $3.0 \times 10^{-7}$ $M$
(b) $1.0 \times 10^{-7}$ $M$
(c) $2.0 \times 10^{-7}$ $M$
(d) $2.1 \times 10^{-7}$ $M$
(e) None of these

For exercises 25–28, consider the titration of 100.0 mL of 1.00 $M$ $H_2A$ ($K_{a_1} = 1.50 \times 10^{-3}$, $K_{a_2} = 2.00 \times 10^{-6}$) with 1.00 $M$ NaOH. Calculate the $[H^+]$:

**25.** Before any NaOH is added:
(a) $3.87 \times 10^{-2}$ $M$
(b) $1.50 \times 10^{-3}$ $M$
(c) $5.48 \times 10^{-5}$ $M$
(d) $2.00 \times 10^{-6}$ $M$
(e) None of these

**26.** After 40.00 mL of 1.00 $M$ NaOH has been added:
(a) $1.50 \times 10^{-3}$ $M$
(b) $1.00 \times 10^{-3}$ $M$
(c) $5.47 \times 10^{-5}$ $M$
(d) $2.25 \times 10^{-3}$ $M$
(e) None of these

**27.** After 100.0 mL of 1.00 $M$ NaOH has been added:
(a) $1.00 \times 10^{-7}$ $M$
(b) $1.41 \times 10^{-10}$ $M$
(c) $1.50 \times 10^{-3}$ $M$
(d) $5.48 \times 10^{-5}$ $M$
(e) None of these

**28.** After 300.0 mL of 1.00 $M$ NaOH has been added:
(a) $4.00 \times 10^{-14}$ $M$
(b) 1.00 $M$
(c) $1.00 \times 10^{-7}$ $M$
(d) $1.41 \times 10^{-10}$ $M$
(e) None of these

# 6 Qualitative Analysis

## CHAPTER OBJECTIVE

To learn to do qualitative analysis by selective precipitation.

## 6.1 Introduction

One of the primary activities in chemistry is the analysis of materials. We want to know what components constitute a substance and how much of each is present. The process of identification of the components is called **qualitative analysis**. The process of determining the amount of each component present is **quantitative analysis**. We will concern ourselves in this chapter with one method of qualitative analysis for a mixture of metal ions in an aqueous solution.

Typically, before we can identify the components of a complex mixture of ions, we must separate them. This can be done in a variety of ways, including **chromatography** and **selective precipitation**. Chromatography involves passing the mixture through a system with a mobile phase and a stationary phase. Those ions in the mixture that show high affinity for the mobile phase move relatively rapidly through the system, while those ions with high affinity for the stationary phase move relatively slowly. Thus the ions spread out (separate) as they pass through the chromatographic system.

Selective precipitation, on the other hand, involves addition of a chemical to the aqueous mixture of ions, which causes precipitation of one ion or a small group of ions. The solid that forms can be separated from the solution by filtration or, more commonly, be centrifuging the solid to the bottom of the test tube and simply pouring off the liquid phase. To understand how selective precipitation works, consider the following example.

**Sample Exercise 6.1**

A solution contains the ions $Ag^+$, $Ba^{2+}$, and $Ni^{2+}$, which are to be separated using the reagents $HCl(aq)$, $H_2SO_4(aq)$, and $NaOH(aq)$. Before we can decide how to use these reagents to separate the ions, we must establish how each reagent reacts with each ion.

**Solution**

First, we take a solution containing $AgNO_3$, add $HCl(aq)$ to it drop by drop and note that a white solid forms as the hydrochloric acid solution mixes with the $AgNO_3$ solution. What is the nature of this solid? To answer this question, it is *essential* to consider the chemical species present in the solution before the solid forms.

Remember that when an ionic solid dissolves in water, it dissociates into its component ions. Thus when solid $AgNO_3$ dissolves in water, the resulting solution contains $Ag^+$ and $NO_3^-$ ions, both surrounded by water molecules (the ions are hydrated). Also, since HCl behaves as a strong acid in water, it is completely dissociated to produce hydrated $H^+$ and $Cl^-$ ions. Thus when the silver nitrate and hydrochloric acid solutions are mixed, the resulting solution contains the ions $H^+$, $Cl^-$, $Ag^+$, and $NO_3^-$.

What reaction occurs that gives the white solid observed in the experiment? Since we know that the solid must be electrically neutral, a cation and an ion must combine to form the solid. The possibilities are

$$H^+(aq) + Cl^-(aq) \xrightarrow{?} HCl(s)$$

$$Ag^+(aq) + NO_3^-(aq) \xrightarrow{?} AgNO_3(s)$$

$$H^+(aq) + NO_3^-(aq) \xrightarrow{?} HNO_3(s)$$

$$Ag^+(aq) + Cl^-(aq) \xrightarrow{?} AgCl(s)$$

We can choose the correct reaction by considering what we know about these substances. Hydrogen chloride is a gas at normal temperatures and pressure; it cannot form a solid under these conditions. Also, we know that the $AgNO_3$ was already dissolved before the solutions were mixed. Adding them together (and thus diluting them) would not cause $AgNO_3$ to precipitate out of solution. The substance $HNO_3$ in aqueous solution is called nitric acid, one of the common strong acids. The compound $HNO_3$ is very soluble in water and does not exist as a solid under these conditions. Thus the only possible identity of the white solid is silver chloride (AgCl). After considering the various alternatives, we can now say that when $AgNO_3(aq)$ and $HCl(aq)$ are mixed, the white solid AgCl precipitates. The appropriate complete ionic equation is

$$Ag^+(aq) + NO_3^-(aq) + H^+(aq) + Cl^-(aq) \longrightarrow AgCl(s) + H^+(aq) + NO_3^-(aq)$$

Note that the $H^+$ and $NO_3^-$ ions do not participate in the chemical change (we call them **spectator ions**) and could be left out. The equation that involves only those ions taking part in the reaction is called the **net ionic equation**. In this case, the net ionic equation is

$$Ag^+(aq) + Cl^-(aq) \longrightarrow AgCl(s)$$

A similar thinking process is needed to analyze the results of the other possible combinations of reagents and cations. For example, when $H_2SO_4(aq)$ is added to the silver nitrate solution, no reaction is noted. That is, no color change occurs, no gas is evolved, no solid forms, and so on. Thus the ions in the mixed solution* ($H^+$, $HSO_4^-$, $SO_4^{2-}$, $Ag^+$, and $NO_3^-$) can coexist there; none of the possible combinations of cations and anions give an insoluble solid.

When the sodium hydroxide solution is added to aqueous silver nitrate, a white solid is seen initially, which rapidly changes to a brown solid. To determine the identity of these solids, we start by considering the various ions in the mixed solution before any reaction has occurred:

$$\underbrace{Ag^+, NO_3^-,}_{AgNO_3(aq)} \quad \underbrace{Na^+, OH^-}_{NaOH(aq)}$$

---

*As noted in section 5.4, $H_2SO_4$ dissociates in two steps. The first step is complete:

$$H_2SO_4(aq) \longrightarrow H^+(aq) + HSO_4^-(aq)$$

and the second step is partial:

$$HSO_4^-(aq) \rightleftarrows H^+(aq) + SO_4^{2-}(aq)$$

Thus, an aqueous solution of $H_2SO_4$ contains $H^+$, $HSO_4^-$, and some $SO_4^{2-}$ ions.

The possible solids are

$$Ag^+(aq) + NO_3^-(aq) \xrightarrow{?} AgNO_3(s)$$
$$Na^+(aq) + OH^-(aq) \xrightarrow{?} NaOH(s)$$
$$Na^+(aq) + NO_3^-(aq) \xrightarrow{?} NaNO_3(s)$$
$$Ag^+(aq) + OH^-(aq) \xrightarrow{?} AgOH(s)$$

The first two can be ruled out because the $AgNO_3$ and NaOH were completely dissolved in the original solutions and the dilution produced by mixing them will make them more soluble, not less. The third reaction, producing solid $NaNO_3$, can be ruled out by noting the solubility rules given in Table 6.1. Ionic substances that contain the nitrate anion are known to be very soluble in water, so $NaNO_3(s)$ will not form under these conditions. Thus the solid formed when NaOH(*aq*) is added to $AgNO_3(aq)$ must be AgOH(*s*), as given by the fourth equation.

There is one more observation to explain: the color change from white to brown. This observation is consistent with a fact known to an experienced chemist: Hydroxide precipitates sometimes spontaneously convert to the corresponding oxide compound by elimination of water. For example, the conversion from silver hydroxide (AgOH) to silver oxide ($Ag_2O$) can be represented by the following equation:

$$\underset{\text{white}}{AgOH} \quad \underset{\text{white}}{HOAg} \longrightarrow \underset{\text{brown}}{Ag_2O} + H_2O$$

where a molecule of water is eliminated between two AgOH groups.

**Table 6.1** General rules for solubility of salts in water

1. Most nitrate ($NO_3^-$) salts are soluble.
2. Most salts of $Na^+$, $K^+$, and $NH_4^+$ are soluble.
3. Most chloride salts are soluble. Notable exceptions are AgCl, $PbCl_2$, and $Hg_2Cl_2$.
4. Most sulfate salts are soluble. Notable exceptions are $BaSO_4$, $PbSO_4$, and $CaSO_4$.
5. Most hydroxide salts are only slightly soluble. The important soluble hydroxides are NaOH, KOH, and $Ca(OH)_2$.
6. Most sulfide ($S^{2-}$), carbonate ($CO_3^{2-}$), and phosphate ($PO_4^{3-}$) salts are only slightly soluble.

We now have accounted for the reactions of aqueous silver nitrate with hydrochloric acid, sulfuric acid, and aqueous sodium hydroxide. The analogous observations when solutions of $Ba(NO_3)_2$ and $Ni(NO_3)_2$ react with these same reagents are summarized in Table 6.2.

You should note that interpreting the observations of the experiments described above requires knowledge of the relevant chemical principles, but it also requires a

**Table 6.2** Summary of the results for $Ag^+$, $Ba^{2+}$, and $Ni^{2+}$ reacting with $HCl(aq)$, $H_2SO_4(aq)$, and $NaOH(aq)$

| | Reagent | | |
|---|---|---|---|
| Cation | $HCl(aq)$ | $H_2SO_4(aq)$ | $NaOH(aq)$ |
| $Ag^+$ | White precipitate ($AgCl$) | No reaction | White precipitate ($AgOH$) ⟶ Brown precipitate ($Ag_2O$) |
| $Ba^{2+}$ | No reaction | White precipitate ($BaSO_4$) | No reaction |
| $Ni^{2+}$ | No reaction | No reaction | Light-green precipitate ($Ni(OH)_2$) |

great deal of factual knowledge. At this point, you probably do not possess much knowledge of chemical details (such as colors and solubilities of specific substances), so you will have to look up many of these facts as you proceed in your study of qualitative analysis.

**Sample Exercise 6.2** Now that we have collected the data summarized in Table 6.2, how would we separate an aqueous mixture of $Ag^+$, $Ba^{2+}$, and $Ni^{2+}$?

**Solution**

Note first that only $Ag^+$ reacts with hydrochloric acid to form an insoluble salt. If we add $HCl(aq)$ to the mixture of ions, $AgCl(s)$ will form, which can be centrifuged and separated from the solution. Thus we have separated $Ag^+$ from the original mixture as summarized by the following diagram.

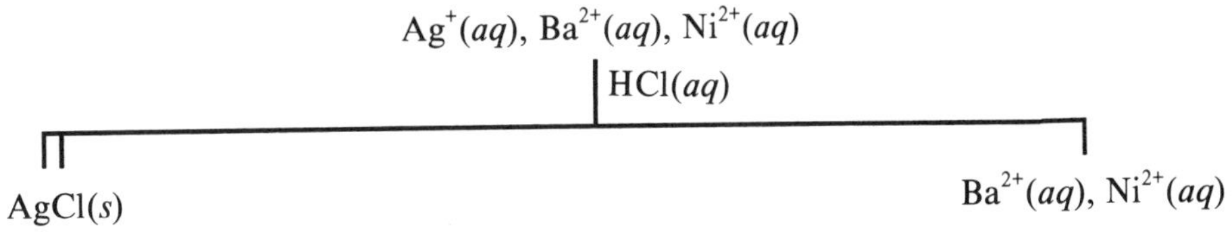

$Ag^+(aq)$, $Ba^{2+}(aq)$, $Ni^{2+}(aq)$

| $HCl(aq)$

$AgCl(s)$ (double line) — $Ba^{2+}(aq)$, $Ni^{2+}(aq)$ (single line)

The double line represents a precipitate, and the single line represents a solution. To separate $Ba^{2+}$ from $Ni^{2+}$, we can add $H_2SO_4(aq)$ to the solution containing these ions, since only $Ba^{2+}$ reacts with sulfuric acid to form an insoluble solid ($BaSO_4$):

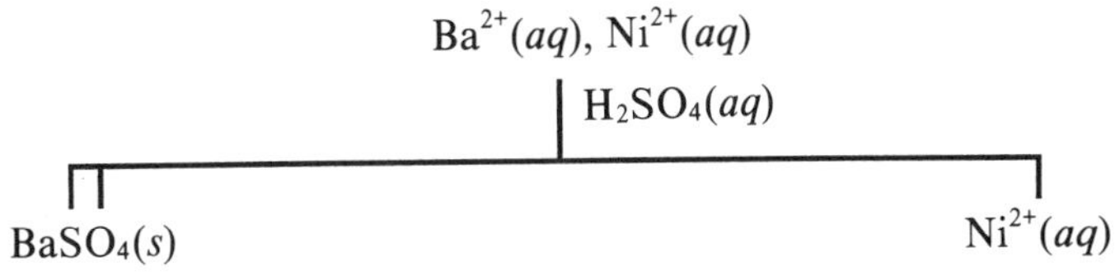

$Ba^{2+}(aq)$, $Ni^{2+}(aq)$

| $H_2SO_4(aq)$

$BaSO_4(s)$ (double line) — $Ni^{2+}(aq)$ (single line)

The solid $BaSO_4$ can be removed, leaving only $Ni^{2+}$ in solution. The nickel ion can be precipitated using $NaOH(aq)$ to form $Ni(OH)_2(s)$. The entire separation scheme can be represented as follows:

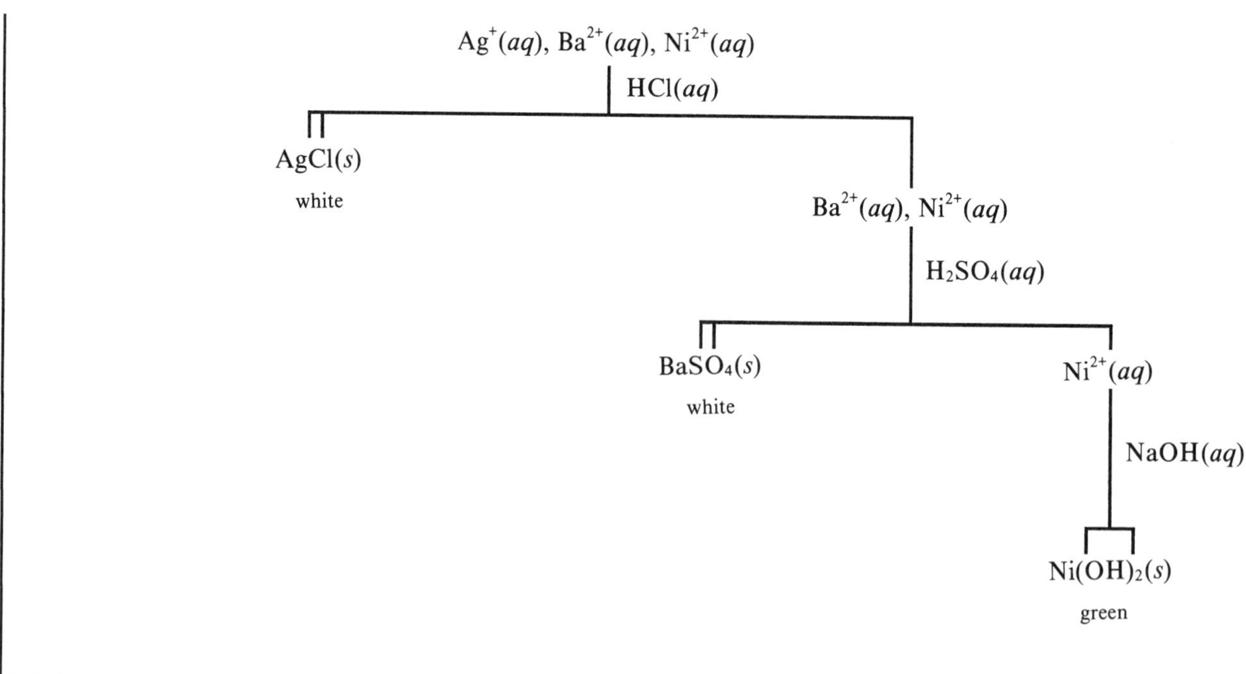


This example has introduced most of the ideas involved in separating ions by selective precipitation. We will now discuss some useful background information before we proceed to more examples.

## 6.2 Solubility Principles

In doing qualitative analysis on a mixture of ions by selective precipitation, the fundamental operation is adding a reagent to form a precipitate. However, we are also interested in the opposite process: how to redissolve a precipitate to proceed with the analysis. These operations can be understood only if we understand solubility equilibria.

Recall from chapter 4 that when a salt MX is placed in contact with water, the dissolution equilibrium

$$MX(s) \rightleftarrows M^+(aq) + X^-(aq)$$

obeys the expression

$$[M^+][X^-] = K_{sp}$$

At equilibrium, the product of the ion concentrations must give the number represented by $K_{sp}$. Thus we can determine whether a precipitate will form when a solution containing $X^-$ is added to a solution containing $M^+$ by comparing the value of the reaction quotient $Q$ (see section 5.5)

$$Q = [M^+]_0[X^-]_0$$

to the value of $K_{sp}$ for MX. If $Q > K_{sp}$, solid MX forms.

What can we do to redissolve MX($s$) once it has formed? We could add more water, but this is impractical since MX($s$) is typically so insoluble in water that a very large amount (many liters) would be needed. The only practical means for redissolving MX($s$) is to add a reagent that increases its solubility. We can increase the solubility of MX($s$), that is, pull the position of the equilibrium

$$MX(s) \rightleftarrows M^+(aq) + X^-(aq)$$

to the right, by introducing a substance that reacts with $M^+$ or $X^-$, or both (see section 5.7). For example, if $X^-$ has a high affinity for $H^+$ (that is, if $X^-$ is an effective base), we can lower the $[X^-]$, and thus increase the solubility of MX, by adding strong acid. This phenomenon can be represented as follows:

$$MX(s) \rightleftarrows M^+(aq) + X^-(aq)$$

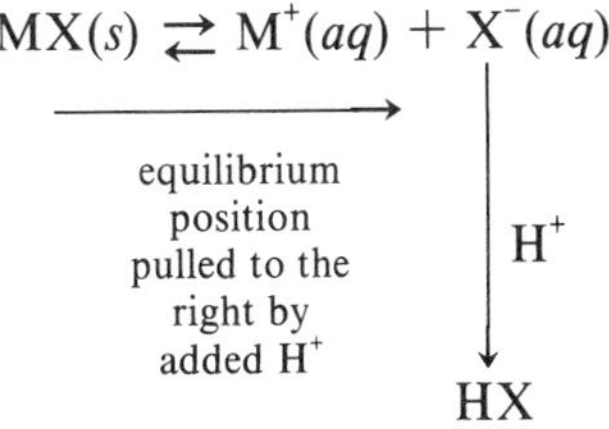


The solubility of a salt can also be increased by adding a ligand to form a complex ion with the metal ion (see sections 5.6 and 5.7):

$$MX(s) \rightleftarrows M^+(aq) + X^-(aq)$$

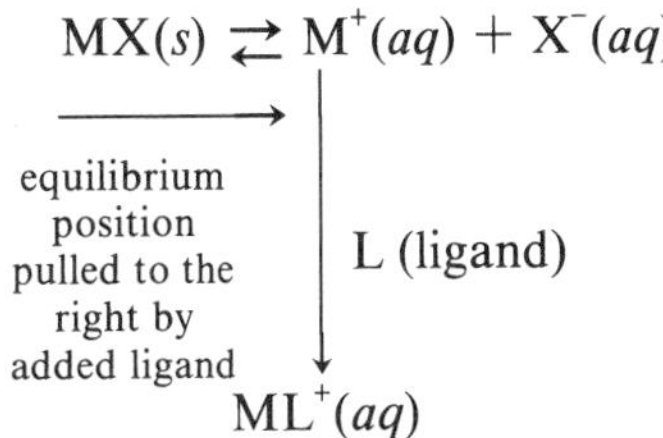


When $H^+$ combines with $X^-$ to form HX and thus lowers the $[X^-]$, more $X^-$ is needed to restore the equilibrium relationship: $[M^+][X^-] = K_{sp}$. The extra $X^-$ is furnished by dissolving more $MX(s)$. Of course, more $M^+$ is also produced in this process. A similar phenomenon occurs when a ligand combines with $M^+$ to lower the $[M^+]$. More $MX(s)$ must dissolve to furnish more $M^+$ (and $X^-$).

The effects of added hydrogen ions and ligands on solubility have very important implications for the qualitative analysis of metal ions by selective precipitation. For example, once an ion has been removed from solution by precipitation, it is often necessary to redissolve the solid to perform further tests. Also, the presence of acid or ligand added in a previous separation step may prevent a desired precipitation from forming. The importance of these ideas will be illustrated in the following exercise.

---

**Sample Exercise 6.3**

Consider the separation of $Ag^+(aq)$ and $Mn^{2+}(aq)$ using the reagents $HCl(aq)$ and $Na_3PO_4(aq)$. The test results for these reagents are shown in Table 6.3. (For practice, identify the precipitates obtained and write the net ionic equations for their formation.)

**Table 6.3** Summary of the results for $Ag^+$ and $Mn^{2+}$ reacting with $HCl(aq)$ and $Na_3PO_4(aq)$

| | | Reagents | |
|---|---|---|---|
| | | $HCl(aq)$ | $Na_3PO_4(aq)$ |
| Ions | $Ag^+$ | White precipitate | Light-yellow precipitate |
| | $Mn^{2+}$ | No reaction | Heavy white precipitate |

#### Solution

Using these results, we can design a scheme in which the $Ag^+$ ion is precipitated by addition of $HCl(aq)$, leaving behind the $Mn^{2+}$ ion, which can then be precipitated by adding $Na_3PO_4(aq)$:

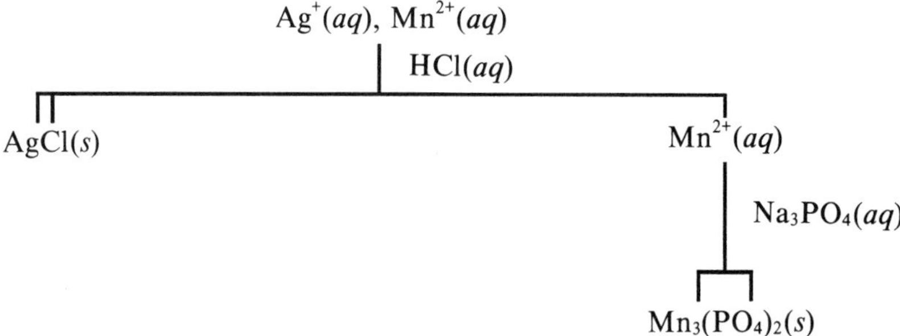


However, when we actually carry out this procedure we are in for a surprise. The AgCl precipitates as expected, but no $Mn_3(PO_4)_2(s)$ forms when several drops of the $Na_3PO_4(aq)$ are added. Why are the results here different from those of the individual tests? The answer involves the pH of the solution. In this scheme, because hydrochloric acid is used as the source of chloride in the first step, $H^+$ is added along with the $Cl^-$ ion. This $H^+$ is still in solution when the $Na_3PO_4$ is added in the second step. Because the $PO_4^{3-}$ ion is a very effective base (see sections 5.4 and 5.7), it reacts with the $H^+$ in solution as follows:

$$H^+(aq) + PO_4^{3-}(aq) \longrightarrow HPO_4^{2-}(aq)$$

$$HPO_4^{2-}(aq) + H^+(aq) \longrightarrow H_2PO_4^-(aq)$$

$$H_2PO_4^-(aq) + H^+(aq) \longrightarrow H_3PO_4(aq)$$

These reactions lower the $[PO_4^{3-}]$ so much that there is not enough of it available to cause precipitation of $Mn_3(PO_4)_2(s)$. There are two ways to overcome this problem. One can continue to add more $Na_3PO_4(aq)$ until the $H^+$ ions are virtually consumed and enough excess phosphate has built up to cause precipitation. Another possibility is to add some NaOH to raise the pH before addition of the $Na_3PO_4(aq)$.

The point of this example is that you must be constantly aware of the acidity of the solution as you perform the steps of a procedure to separate ions by selective precipitation. *Salts with basic anions are soluble in acidic solutions.* That is, you cannot expect to precipitate a solid containing a basic anion from an acidic solution.

On the other hand, if the pH is too high (relatively large $[OH^-]$), the hydroxide compound may precipitate instead of the intended salt. For example, in the above procedure for separating $Ag^+$ and $Mn^{2+}$, if too much $NaOH(aq)$ is added after the hydrochloric acid step, $Mn(OH)_2(s)$ may form. Since most metal hydroxide compounds are quite insoluble, you must always consider the possibility of metal hydroxide precipitation when raising the pH by adding NaOH.

## 6.3 Precipitation with Basic Anions

As we have said in section 6.2, some anions are quite effective bases (have a high affinity for protons). Two examples are the phosphate ion ($PO_4^{3-}$) and the carbonate ion ($CO_3^{2-}$). When such basic anions are placed in water, they react with water to extract a proton:

$$PO_4^{3-}(aq) + H_2O(l) \rightleftarrows HPO_4^{2-}(aq) + OH^-(aq)$$

$$CO_3^{2-}(aq) + H_2O(l) \rightleftarrows HCO_3^-(aq) + OH^-(aq)$$

This means that, for example, in an aqueous solution of $Na_3PO_4$, there are several anions present that can form precipitates with metal ions, namely $PO_4^{3-}$, $OH^-$, and $HPO_4^{2-}$. Since the salts of $HPO_4^{2-}$ are typically quite soluble, the two anions we must be concerned about are $OH^-$ and $PO_4^{3-}$. Thus, whenever $Na_3PO_4(aq)$ is added to a solution containing a metal ion and a precipitate forms, that precipitate is most likely the hydroxide or the phosphate compound. One useful clue that will help you decide which solid has formed is that hydroxide compounds usually produce a gelatinous-looking (jellylike) solid. (Be sure to notice the consistency of the compounds formed in the lab when $NaOH(aq)$ is added to metal ions such as $Ni^{2+}$, $Al^{3+}$, $Cu^{2+}$ and $Co^{2+}$). On the other hand, phosphate salts are usually flocculent* solids composed of large particles. For example, when $Na_3PO_4(aq)$ is added to a solution of $Mn^{2+}$, a flocculent, white precipitate is formed. This indicates that $Mn_3(PO_4)_2$ is formed rather than $Mn(OH)_2$. But when $Na_3PO_4(aq)$ is added to a solution containing $Al^{3+}(aq)$, a white, gelatinous solid forms. This solid is certainly $Al(OH)_3$ rather than $AlPO_4$.

The same principles apply to the addition of $Na_2CO_3(aq)$ to metal ions in solution. In fact, for $Na_2CO_3(aq)$ the precipitate most often formed is the hydroxide compound rather than the carbonate.

## 6.4 Formation of Basic Salts

In section 6.3 we saw that when basic anions are added to solutions of metal ions, the resulting precipitate can be either the hydroxide compound or the compound containing the basic anion. In rare cases the precipitate can be a combination of these. This phenomenon is most often observed for metal ions that behave as weak acids in aqueous solution (see section 2.8). For example, $Bi^{3+}(aq)$, which represents the complex ion $Bi(H_2O)_6^{3+}$, acts as a weak acid as follows:

$$Bi(H_2O)_6^{3+}(aq) + H_2O(l) \rightleftarrows Bi(OH)(H_2O)_5^{2+}(aq) + H_3O^+(aq)$$

This means that when an anion is added to a solution containing $Bi^{3+}(aq)$, there are at least two cations ($Bi(H_2O)_6^{3+}$ and $Bi(OH)(H_2O)_5^{2+}$) that could form a solid salt. In fact, when 3 $M$ HCl is added to a solution containing $Bi^{3+}(aq)$, no reaction is noted, but when this mixture is then diluted (water is added), a flocculent, white precipitate is formed. This can be explained as follows. When 3 $M$ HCl is added, the position of the equilibrium

$$Bi(H_2O)_6^{3+}(aq) + H_2O(l) \rightleftarrows Bi(OH)(H_2O)_5^{2+}(aq) + H_3O^+(aq)$$

is shifted to the left by the added $H^+(H_3O^+)$. Apparently, chloride does not form an insoluble salt with $Bi(H_2O)_6^{3+}$ (no precipitate is observed when 3 $M$ HCl is added). However, when water is subsequently added, the $[H_3O^+]$ is lowered and the position of the above equilibrium shifts to the right, increasing the concentration of $Bi(OH)(H_2O)_5^{2+}$. The precipitate that forms must be a combination of $Cl^-$ and $Bi(OH)(H_2O)_5^{2+}$. Leaving out the water molecules, we might suspect the formula of this solid to be $Bi(OH)Cl_2$. However, the compound loses a hydrogen ion and becomes the corresponding oxy salt, BiOCl. The bismuth ion also exhibits similar behavior with other anions. In general, the aqueous chemistry of the bismuth ion is very complicated and requires careful interpretation.

## 6.5 Amphoteric Hydroxide Compounds

We have seen (section 2.8) that some hydrated metal ions can behave as acids because the metal ion causes the O—H bond to be more easily broken to produce $H^+$. Some metal hydroxides also show this behavior. For example, aluminum hydroxide, $Al(OH)_3$, which is quite insoluble in water ($K_{sp} \approx 10^{-32}$) is soluble in both acidic and basic solutions. Since $Al(OH)_3(s)$ contains the $OH^-$ anion, it is expected to be soluble

*Flocculent means fluffy; resembling tufts of wool.

in acids, since the $H^+$ in solution reacts with $OH^-$ to form water and "pulls" the $Al(OH)_3(s)$ into solution:

$$Al(OH)_3(s) \rightleftarrows Al^{3+}(aq) + 3OH^-(aq)$$

equilibrium pulled to the right by $H^+$ reacting with $OH^-$ (arrow →); $OH^-$ + $H^+(aq)$ ↓ $H_2O(l)$

By this mechanism all hydroxide compounds are more soluble in acidic solution than in pure water.

In the above process the aluminum hydroxide is acting as a base (furnishing $OH^-$ ions that react with $H^+$ ions). However, aluminum hydroxide can also function as an acid. This follows from the ability of the $Al^{3+}$ ion to weaken the O—H bond. Thus $Al(OH)_3$ reacts with $OH^-$ as follows:

$$Al(OH)_3(s) + OH^-(aq) \longrightarrow AlO_2^-(aq) + 2H_2O(l)$$

causing the solid $Al(OH)_3$ to dissolve.

Thus $Al(OH)_3(s)$ can function as a base or as an acid (it is said to be amphoteric) and dissolves in both acidic and basic solutions. There are many other metal hydroxides that show this type of behavior, such as, $Cu(OH)_2$, $Pb(OH)_2$, and $Zn(OH)_2$. Along with $Al(OH)_3$ these hydroxides are all soluble in concentrated $NaOH(aq)$.

We can use the amphoteric nature of these hydroxides to separate them from nonamphoteric hydroxide compounds using concentrated $NaOH(aq)$. For example, if the mixture $Fe^{3+}$, $Ni^{2+}$, $Al^{3+}$, and $Cu^{2+}$ is treated with concentrated $NaOH(aq)$, the following results are obtained:

$Fe^{3+}$, $Ni^{2+}$, $Al^{3+}$, $Cu^{2+}$

concentrated $NaOH(aq)$

| | |
|---|---|
| $Fe(OH)_3(s)$<br>$Ni(OH)_2(s)$ | $AlO_2^-(aq)$<br>$CuO_2^{2-}(aq)$ |

In this way, we can separate the original mixture of four ions into groups of two.

## 6.6 Precipitation with Aqueous Ammonia

As discussed in section 2.6, when ammonia is dissolved in water the following equilibrium is established:

$$NH_3(aq) + H_2O(l) \rightleftarrows NH_4^+(aq) + OH^-(aq) \qquad K_b = 1.8 \times 10^{-5}$$

Since $K_b$ is small, only about 1% of the original ammonia molecules exist as $NH_4^+$ ions. Thus in a 15 $M$ $NH_3$ solution the actual concentration of $NH_3$ is only very slightly less than 15 $M$. The concentration of $OH^-$ in this solution is $1.6 \times 10^{-2}$ $M$. To interpret the results when aqueous ammonia is added to a metal ion in solution, we must consider both the $NH_3$ molecules, which behave as ligands toward metal ions, and the $OH^-$ ions.

**Sample Exercise 6.4**

When 15 $M$ $NH_3(aq)$ is added drop by drop to an aqueous solution containing $Cu^{2+}$, initially a blue gelatinous precipitate forms, which then dissolves on continued addition of ammonia to give a deep blue solution. How can we account for these observations?

**Solution**

The solid that forms is gelatinous, a characteristic of hydroxide solids. Thus it is reasonable to suggest that the blue solid is $Cu(OH)_2$, which forms by the reaction between the $Cu^{2+}$ ions and the $OH^-$ ions in the ammonia solution. The $Cu(OH)_2(s)$ dissolves as more ammonia is added, because the $NH_3$ molecules react with the solid to form the deep blue complex ion $Cu(NH_3)_4^{2+}$. This process can be represented as follows:

$$Cu(OH)_2(s) \rightleftarrows Cu^{2+}(aq) + 2OH^-(aq)$$

$$\downarrow \text{excess } NH_3(aq)$$

$$Cu(NH_3)_4^{2+}(aq)$$

The formation of $Cu(NH_3)_4^{2+}$ lowers the concentration of $Cu^{2+}$, pulling the solubility equilibrium to the right and causing the $Cu(OH)_2(s)$ to dissolve.

The main point of this example is that when you interpret the reactions involving aqueous ammonia and metal ions, you must consider both the $OH^-$ ions and the $NH_3$ molecules present in the ammonia solution.

## 6.7 Precipitation with Sulfide

An anion commonly used in selective precipitation procedures is sulfide, $S^{2-}$. One of the main advantages of the sulfide ion is that as a base, its concentration can be controlled by the acidity of the solution (see section 4.6). Although sulfide can be introduced into the solution in several ways, the pertinent equilibria always involve $H_2S$, $HS^-$, and $S^{2-}$:

$$H_2S(aq) \rightleftarrows H^+(aq) + HS^-(aq) \qquad K_{a_1} = 1.0 \times 10^{-7}$$

$$HS^-(aq) \rightleftarrows H^+(aq) + S^{2-}(aq) \qquad K_{a_2} = 1.3 \times 10^{-13}$$

Note from these equilibria that if the solution is acidic (large $[H^+]$), the equilibrium position will be pushed to the left and $[S^{2-}]$ will be small. On the other hand, in basic solution (small $[H^+]$), the equilibria will be pulled to the right to produce a relatively large $[S^{2-}]$. Because of the basicity of the sulfide ion and because the $K_{sp}$ values for sulfide salts vary widely, the sulfide ion is very useful for selective precipitation of metal ions.

**Sample Exercise 6.5**

Consider a mixture containing $1.0 \times 10^{-3}$ $M$ $Cu^{2+}$ and $1.0 \times 10^{-3}$ $M$ $Mn^{2+}$. Since the $K_{sp}$ for CuS is very small ($8.5 \times 10^{-45}$) compared to that for MnS ($2.3 \times 10^{-13}$), the two ions can be separated by providing a sulfide ion concentration in solution that will precipitate CuS but not MnS. How can we choose a pH value to accomplish this separation?

**Solution**

First, we assume that we have introduced sufficient $H_2S$ to saturate the solution. This will give an $H_2S$ concentration of 0.1 $M$. Now we must determine a value of $[S^{2-}]$ that will cause precipitation of CuS but not MnS. We can obtain this information from the $K_{sp}$ expressions:

For $CuS(s)$:

$$CuS(s) \rightleftarrows Cu^{2+}(aq) + S^{2-}(aq)$$

$$K_{sp} = [Cu^{2+}][S^{2-}] = 8.5 \times 10^{-45}$$

In the mixture, $[Cu^{2+}] = 1.0 \times 10^{-3}\ M$
Thus

$$\frac{K_{sp}}{[Cu^{2+}]} = \frac{8.5 \times 10^{-45}}{1.0 \times 10^{-3}} = 8.5 \times 10^{-42}\ M = [S^{2-}]$$

This $[S^{2-}]$ represents the concentration that we must *exceed* to cause precipitation of CuS, since if $[S^{2-}] > 8.5 \times 10^{-42}\ M$, the reaction quotient $Q$ will be greater than the $K_{sp}$ for CuS.

For $MnS(s)$:

$$MnS(s) \rightleftarrows Mn^{2+}(aq) + S^{2-}(aq)$$

$$K_{sp} = [Mn^{2+}][S^{2-}] = 2.3 \times 10^{-13}$$

In the mixture,

$$[Mn^{2+}] = 1.0 \times 10^{-3}\ M$$

Proceeding as before,

$$\frac{K_{sp}}{[Mn^{2+}]} = \frac{2.3 \times 10^{-13}}{1.0 \times 10^{-3}} = 2.3 \times 10^{-10}\ M = [S^{2-}]$$

This value of $[S^{2-}]$ represents the largest concentration of sulfide that can be present *without* causing precipitation of MnS. That is, for this value of $[S^{2-}]$, $Q = K_{sp}$ and no precipitation of MnS occurs. However, for any $[S^{2-}] > 2.3 \times 10^{-10}\ M$, $MnS(s)$ will form.

These results tell us that to precipitate CuS but not MnS we must have

$$[S^{2-}] > 8.5 \times 10^{-45}\ M$$

to precipitate CuS, but

$$[S^{2-}] < 2.3 \times 10^{-10}\ M$$

to prevent precipitation of MnS. Let's choose $[S^{2-}] = 1.0 \times 10^{-10}\ M$, since this will precipitate virtually all of the $Cu^{2+}$ as CuS but is well below the limit for MnS.

We can choose the appropriate pH by using the combined dissociation equilibria for $H_2S$. If we multiply together the two equilibrium expressions for $H_2S$, we have

$$\frac{[H^+][HS^-]}{[H_2S]} \times \frac{[H^+][S^{2-}]}{[HS^-]} = K_{a_1} \times K_{a_2}$$

which gives

$$\frac{[H^+]^2[S^{2-}]}{[H_2S]} = K_{a_1} \times K_{a_2} = (1.0 \times 10^{-7}) \times (1.3 \times 10^{-13})$$

$$= 1.3 \times 10^{-20}$$

In the present case, we know that the solution is saturated with $H_2S$ ($[H_2S] = 0.1\ M$) and that the sulfide ion concentration needed is $1.0 \times 10^{-10}\ M$. Substituting these values gives

$$\frac{[H^+]^2[S^{2-}]}{[H_2S]} = \frac{[H^+]^2(1.0 \times 10^{-10})}{0.1} = 1.3 \times 10^{-20}$$

Solving for $[H^+]^2$, we have

$$[H^+]^2 = \frac{(0.1)(1.3 \times 10^{-20})}{1.0 \times 10^{-10}}$$

$$[H^+] = 4 \times 10^{-6}\ M$$

The pH value needed is

$$pH = -\log [H^+] = 5.4$$

To summarize: for a mixture containing $10^{-3}$ *M* $Cu^{2+}$ and $10^{-3}$ *M* $Mn^{2+}$, we can precipitate CuS while leaving the $Mn^{2+}$ in solution by maintaining a pH of 5.4 and saturating the solution with $H_2S$.

We can roughly divide the sulfide salts of the various metal ions into two classes: very insoluble and relatively soluble. From the previous example, CuS ($K_{sp} = 8.5 \times 10^{-45}$) resides in the first class and MnS ($K_{sp} = 2.3 \times 10^{-13}$) in the second. In qualitative analysis schemes, a large mixture of metal ions is often treated with $H_2S$ to separate it into smaller groups. If $H_2S$ is added to the mixture at an acidic pH, only the very insoluble sulfides will precipitate. If the solution is then made more basic, the more soluble sulfide salts will precipitate. An example is illustrated in the following diagram:

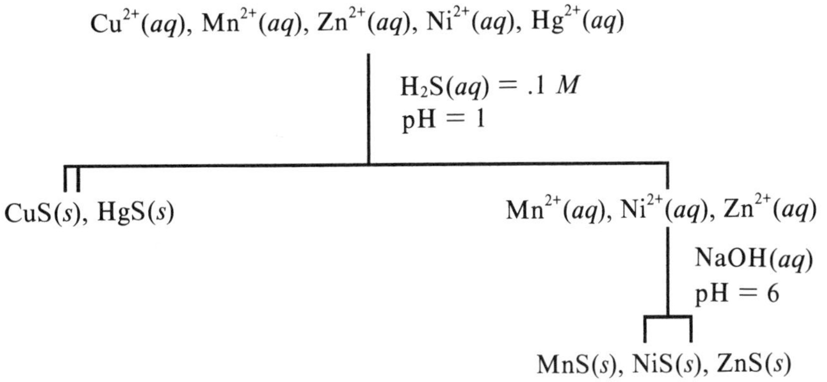

$Cu^{2+}(aq)$, $Mn^{2+}(aq)$, $Zn^{2+}(aq)$, $Ni^{2+}(aq)$, $Hg^{2+}(aq)$

$H_2S(aq) = .1\ M$
pH = 1

$CuS(s)$, $HgS(s)$ | $Mn^{2+}(aq)$, $Ni^{2+}(aq)$, $Zn^{2+}(aq)$

$NaOH(aq)$
pH = 6

$MnS(s)$, $NiS(s)$, $ZnS(s)$

In adding $NaOH(aq)$ to raise the pH, you must keep in mind that most metal ions form insoluble hydroxide compounds. Thus the hydroxide rather than the sulfide may be obtained as the precipitate for a given metal ion, but this represents no problem, since the separation is achieved either way.

As mentioned earlier, $H_2S$ can be introduced into a solution in several ways, but the most convenient method involves reaction of water with thioacetamide, $CH_3CSNH_2$. In acidic solution this reaction is

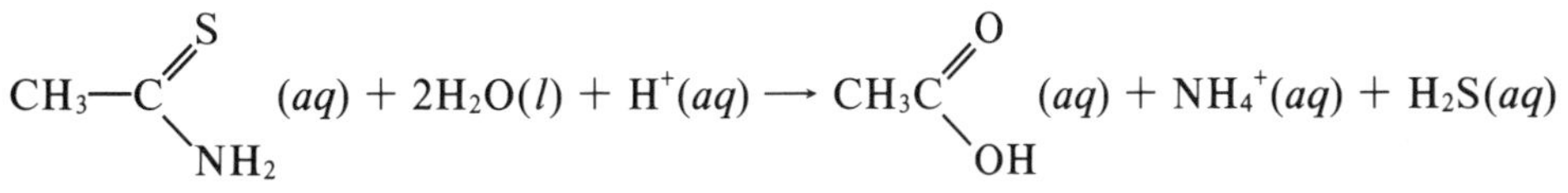

$$\underset{\text{thioacetamide}}{CH_3-C(=S)NH_2}(aq) + 2H_2O(l) + H^+(aq) \longrightarrow \underset{\text{acetic acid}}{CH_3C(=O)OH}(aq) + NH_4^+(aq) + H_2S(aq)$$

The thioacetamide is usually introduced by adding about ten drops of an 8% ($\approx$1 *M*) aqueous solution of thioacetamide to $\approx$1 mL of the solution containing the cations being analyzed. Because the reaction of thioacetamide with water is relatively slow at 25°C, the solution should be heated for about five minutes in a boiling water bath after addition of the thioacetamide.

Because hydrogen sulfide is toxic and has a foul odor, all operations involving $H_2S$ should be carried out in a fume hood.

## 6.8 Oxidation-Reduction Reactions

Oxidation-reduction reactions, those that involve a transfer of electrons, sometimes occur in connection with the qualitative analysis of cations. For example, nitric acid is often used to dissolve acid-soluble solids. Because nitric acid is a relatively strong oxidizing agent, oxidation of the cation and/or the anion may occur in this situation. An important case in point occurs when metal sulfide compounds are dissolved with nitric acid. Because nitric acid can oxidize sulfide to sulfur ($S^{2-} \longrightarrow S + 2e^-$), an off-white or tan suspension of sulfur (milky-looking) often occurs on treating sulfides with nitric acid. Some metal ions can also oxidize the sulfide ion. An example is $Fe^{3+}$, which can oxidize $S^{2-}$ to S by changing to $Fe^{2+}$:

$$2Fe^{3+}(aq) + S^{2-}(aq) \longrightarrow 2Fe^{2+}(aq) + S(s)$$

In the presence of excess sulfide ion, the $Fe^{2+}$ will precipitate as FeS under appropriate pH conditions.

Many metal ions can be oxidized by exposure to the oxygen in air. For example, $Fe^{2+}$ is easily oxidized to $Fe^{3+}$ by $O_2$. Thus a mixture that initially contains $Fe^{2+}$ may contain significant amounts of $Fe^{3+}$ after it is handled in the air.

Another similar case involves the plus one ion of mercury, which usually exists as the dimer $Hg_2^{2+}$. This ion is readily oxidized to $Hg^{2+}$ by oxygen. The Hg(I) ion can also disproportionate (undergo both oxidation and reduction) to Hg metal and the $Hg^{2+}$ ion:

$$Hg_2^{2+}(aq) \longrightarrow Hg(l) + Hg^{2+}(aq)$$

This occurs when $Hg_2Cl_2(s)$ is treated with aqueous ammonia:

$$Hg_2Cl_2(s) + NH_3(aq) \longrightarrow Hg(l) + HgNH_2Cl(s) + Cl^-(aq)$$

where $HgNH_2Cl$ is a white compound that contains the $Hg^{2+}$ cation and the $Cl^-$ and $NH_2^-$ anions. The mercury that forms is very finely divided and has a black color. Mixed together, the Hg and $HgNH_2Cl$ produce a gray-appearing solid.

Another oxidation-reduction process useful in qualitative analysis involves the oxidation of $Mn^{2+}$ to $MnO_4^-$ by a strong oxidizing agent such as sodium bismuthate ($NaBiO_3$):

$$\underset{\text{almost colorless}}{2Mn^{2+}(aq)} + 5BiO_3^-(aq) + 14H^+(aq) \longrightarrow \underset{\text{intense purple}}{2MnO_4^-(aq)} + 5Bi^{3+}(aq) + 7H_2O(l)$$

This reaction is particularly useful for the identification of $Mn^{2+}$, since under these conditions it forms the distinctively purple permanganate ion ($MnO_4^-$).

## 6.9 The Effect of Temperature on Solubility

Most ionic solids become more soluble as the temperature is increased. One case for which this characteristic is particularly useful is lead chloride, which can be precipitated in ice cold water and dissolved in boiling water. The following example shows how this fact can be used to separate $Pb^{2+}$ from $Ag^+$ and $Hg_2^{2+}$. If cold HCl(*aq*) is added to a mixture of $Ag^+$, $Pb^{2+}$ and $Hg_2^{2+}$, then AgCl, $Hg_2Cl_2$, and $PbCl_2$ will precipitate. The solid chloride salts are then centrifuged and the excess solution poured off. Approximately one milliliter of distilled water is then added to the test tube containing the solid and this tube is placed in a bath of boiling water. Stirring the solid for a few minutes will cause the lead chloride to dissolve in the hot water, which then can be poured off. This allows the lead to be separated from the other ions as summarized in the following diagram:

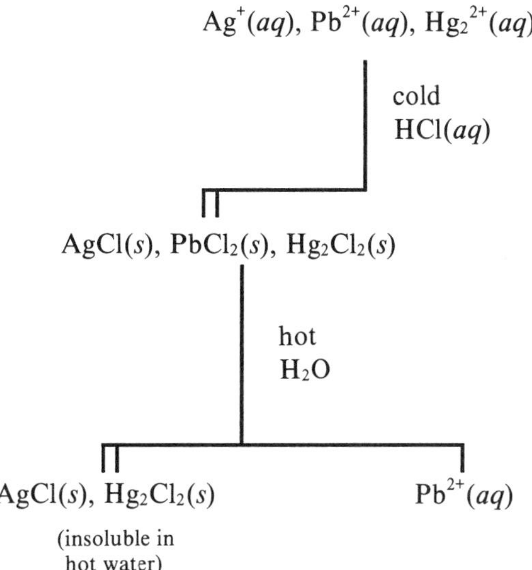


## 6.10 A Qualitative Analysis Experiment

### Introduction

In this section, we discuss a qualitative analysis experiment that is exploratory in nature rather than highly structured. The point of this experiment is to allow you to collect data on the chemical behavior of several metal ions with a set of specified reagents and then to design your own separation and identification schemes.

This experiment is supposed to work like the scientific method works: observations are made, hypotheses are put forth, and experiments are done to test these hypotheses. You will first gather data on how the ions and reagents react, then design a scheme for separation and identification of the ions, and then test your scheme by using it to analyze a *known* mixture of ions. You will probably find that the first scheme you design doesn't work. If this happens, use your observations and the principles discussed earlier in this chapter to modify your scheme. The difficulties usually arise from chemicals added in earlier steps.

One very common problem is failure to maintain an appropriate value of pH. For example, you cannot expect to form an acid-soluble precipitate in a solution that is strongly acidic. Another problem that can sometimes occur results from addition in a previous step of molecules that can behave as ligands. For example, one could not expect to form $AgCl(s)$ by adding $NaCl(aq)$ to a solution containing $Ag^{+}$ and $NH_3$. In this case, most of the silver ions would be present as complex ions ($Ag(NH_3)_2^{+}$) rather than as $Ag^{+}$, and too few $Ag^{+}$ ions would be present to form the AgCl precipitate.

As you do this experiment, keep in mind that science typically does not proceed smoothly. Rather, many blind alleys are explored before a solution is found. In science, trial and error is often necessary. When you are having trouble separating two ions, make a solution containing those two ions and try various ways of achieving the separation. Rely heavily on your own observations as you do this experiment. In trying to interpret your results, it is absolutely crucial that you think about the species that are floating around in the solution. Use these species to postulate the various alternatives and then use your observations and your knowledge of the facts and theories of chemistry to figure out what is going on. Be patient. Remember that you will probably learn the most when your scheme fails and you have to determine how to overcome the problem. This experiment can be frustrating at times because it doesn't give you a recipe for every situation, but you should learn more chemistry and ultimately find this approach more satisfying than a cookbook approach:

You will study eleven cations in this experiment:

$Pb^{2+}$, $Ba^{2+}$, $Bi^{3+}$, $Fe^{3+}$, $Hg^{2+}$, $Mn^{2+}$, $Ag^{+}$, $Cu^{2+}$, $Ni^{2+}$, $Zn^{2+}$, $Al^{3+}$

First you will run tests with various reagents on each ion to determine its unique chemical properties. No two ions will behave exactly the same way, and in a sense, you will determine a chemical fingerprint for each one. As you try to interpret your observations from each test, you should keep three questions in mind:

1. What species are present in solution before any reaction takes place?
2. What possible reactions can occur between these species?
3. Which of the possible reactions explains the observations?

The discussions in the previous sections should help you answer these questions, but you may also have to consult your course textbook, a book on qualitative analysis (see the references at the end of this chapter), or a standard reference book such as the *Handbook of Chemistry and Physics.*

## Determination of the Chemical Properties of the Cations

*Reactions with Chloride, Sulfate, Phosphate, Carbonate, and Chromate Ions* The reagents to be used are 3.0 *M* solutions of HCl and $H_2SO_4$ and 0.1 *M* solutions of $Na_3PO_4$, $Na_2CO_3$, and $K_2CrO_4$. Test each of the eleven cations with the five reagents and record the results in grid form in your lab notebook. List the test reagents horizontally and the cations vertically. Leave enough room in each grid space to describe your observations, such as "no observable change" (N.C.); "precipitate forms" (note the color and nature of the precipitate *and* whether it is soluble in 6.0 *M* nitric acid); "solution changes color"; or any other change that occurs.

To carry out each test, place about 1 mL of the .1 *M* cation solution in a small test tube and slowly add the test reagent, one drop at a time. Mix the solution by tapping the test tube with your finger. If a precipitate forms, let it settle and carefully pour off the liquid. Add 6.0 *M* nitric acid, drop by drop, to the solid, with mixing, to see if the precipitate dissolves. Be certain to indicate the behavior of the precipitate toward nitric acid. When you test $Bi^{3+}$ with HCl, after the addition of acid, dilute the solution with deionized or distilled water, and note the result.

*Reactions with Aqueous Solutions of Ammonia and Sodium Hydroxide* Each cation is to be tested with 3 *M* ammonia, 15 *M* ammonia, 3 *M* sodium hydroxide, and 19 *M* sodium hydroxide (50% by mass).

To approximately 1 mL of cation solution add, drop by drop, a *few* drops of 3 *M* sodium hydroxide, and note the results. Then add about 1 mL of 19 *M* sodium hydroxide, mix well, and note the results. Carry out the same tests, in the same sequence, using 3 *M* and 15 *M* ammonia. Note the results, after each addition, in grid form in the lab notebook. Make special note of amphoteric behavior or complex ion formation. The results of the ammonia and sodium hydroxide tests will be especially important later on, when you are working on the unknowns, so make careful observations.

*Reactions with Sulfide Ion* Knowing whether or not a cation forms a sulfide precipitate under certain conditions will be important in designing separation schemes. First, test each cation for sulfide precipitate formation in acidic solution. To about 1 mL of the cation solution add *one* drop of 2 *M* nitric acid. Then saturate the solution with $H_2S$. This is most conveniently done by adding about ten drops of 8% (by mass) thioacetamide solution and heating the resulting mixture in a boiling water bath for five minutes. If a precipitate forms, centrifuge, decant the liquid, and test the precipitate to see if it is soluble in 1 mL of 6 *M* nitric acid by heating the acid-precipitate mixture in a boiling water bath with occasional stirring. After five minutes or so, you may notice the precipitate has dissolved but a yellowish substance is floating on the solution. This is elemental sulfur, produced by the oxidation of sulfide ion by nitric acid. Record all observations in grid form in your lab notebook.

Now repeat the procedure, but under basic conditions. To 1 mL of cation solution add 1 mL of 3 *M* ammonia and 1 mL of 3 *M* ammonium chloride. This buffers the solution at about pH 9. Now add 1 mL of 8% thioacetamide and heat the mixture in a boiling water bath for five minutes. If a precipitate forms, certrifuge, decant the liquid, and then add 1 mL of 6 *M* nitric acid, as above, to test for solubility. Record all observations in the lab notebook.

## The Unknowns

Once you have completed the individual ion tests, obtain from your instructor a list of the possible cations that may be in your first unknown. Then, using your knowledge of the chemical properties of the cations, devise a detailed scheme for the separation and identification of this group of cations. Be sure to include equations for the reactions that occur at each step in the scheme.

After your analysis scheme has been checked by the instructor, test it by preparing a known sample containing all of the cations assigned to you and subjecting this to analysis. When *you* are satisfied with your scheme, obtain an unknown sample from the instructor. Each unknown will contain some or all of the ions on your list.

## Some Confirmatory Tests for the Cations

It is often useful to confirm the presence of a cation in an unknown sample using reactions that form characteristic precipitates or complex ions. Before an ion is tested to confirm its presence, it should be separated from the other cations. Before using a confirmatory test for an unknown ion, the test should be carried out with a known sample so that you will recognize a positive result.

Table 6.4 lists tests that can be used to confirm the presence of an ion. Others can be found in texts dealing with qualitative analysis.

**Table 6.4** Test to confirm presence of an ion

| Ion | Test | Products and Observations |
|---|---|---|
| $Fe^{3+}$ | 1. Add small amount of solid KSCN to mildly acidic solution of cation<br>2. Add $K_4[Fe(CN)_6]$ solution to mildly acidic solution of ion. | 1. $Fe(SCN)^{2+}$, blood-red solution<br>2. $KFe[Fe(CN)_6]$, royal-blue precipitate |
| $Mn^{2+}$ | Add solid $NaBiO_3$ and five drops of 6 *M* $HNO_3$, stir, centrifuge | $MnO_4^-$, purple solution |
| $Pb^{2+}$ | Add small amount of solid KI to solution of ion | $PbI_2$, yellow precipitate first forms, then dissolves in excess KI, forming $PbI_4^{2-}$, colorless |
| $Cu^{2+}$ | 1. Add solid KI to ion solution<br>2. Add $K_4[Fe(CN)_6]$ to mildly acidic solution of ion | 1. Orange-brown solution of $I_2$ + tan precipitate<br>2. $Cu_2[Fe(CN)_6]$, red-brown precipitate |
| $Ni^{2+}$ | Add dimethylglyoxime (DMG) reagent to ion solution made mildly basic with $NH_3$ | $Ni(DMG)_2$, voluminous strawberry-red precipitate |
| $Bi^{3+}$ | Add solid KI, and follow after several seconds with a water dilution | First a green-black precipitate of $BiI_3$ forms; later an orange solution of $BiI_4^-$; dilution regenerates $BiI_3$ precipitate |
| $Ag^+$ | Add solid KI to solution of ion | AgI, light-yellow precipitate forms. |

| | | |
|---|---|---|
| $Al^{3+}$ | Form $Al(OH)_3$ precipitate with $NH_3$, wash with hot water, centrifuge, dissolve in three to five drops of 3 *M* $HNO_3$, add two drops of aluminon reagent, make just basic with ammonia, centrifuge | $Al(OH)_3$ precipitate, dyed red due to adsorption of aluminon—test must be carefully done |
| $Zn^{2+}$ | 1. Basic thioacetamide | 1. ZnS, white precipitate soluble in HCl but not in acetic acid |
| | 2. Add $K_4[Fe(CN)_6]$ to solution of ion made acidic with acetic acid | 2. $K_2Zn_3[Fe(CN)_6]_2$ gray-white precipitate |
| $Hg^{2+}$ | Add excess 6 *M* NaOH | HgO, yellow precipitate soluble in acid |
| $Ba^{2+}$ | Add $K_2CrO_4$ to solution of ion | $BaCrO_4$, yellow precipitate |

## Miscellaneous Notes

*Safety* As always when working in the chemical laboratory, you must observe proper safety precautions. Wear appropriate eye protection at all times and be cautious as you handle the reagents. Be especially careful in handling the strong acids (hydrochloric, nitric, and sulfuric) and the sodium hydroxide solutions. These corrosive chemicals will cause severe burns. If you spill a reagent on your skin, immediately wash the area with large quantities of water and tell your instructor.

Hydrogen sulfide is very toxic and must be handled with great care (see below).

*Use of Hydrogen Sulfide* Hydrogen sulfide is a highly toxic, foul-smelling gas that should be used in a fume hood with caution. Sometimes it is necessary to purge a solution of dissolved sulfide to prevent interference with successive tests. This can be accomplished by adding nitric acid (to react with $S^{2-}$ to form volatile $H_2S$ and to oxidize $S^{2-}$ to S) and carefully boiling in a hood for a few minutes in an evaporating dish.

*Rinsing Precipitates* After a precipitate has formed and has been centrifuged, and after the remaining solution has been poured off (decanted), it is important to rinse the precipitate with deionized or distilled water. This procedure washes away the contaminating ions that adhere to the surface of the precipitate and that can cause errors in subsequent tests.

Wash the precipitate by adding approximately 1 mL of distilled or deionized water to the isolated solid in a test tube (the liquid has been decanted), stir, centrifuge, and pour off the liquid. This should be done a couple of times.

*Hints*

1. Sometimes students have trouble separating $Cu^{2+}$ and $Zn^{2+}$. If you are having problems with these ions, try each in a saturated $H_2S$ solution acidified with HCl. The results should help you design a separation scheme.
2. $Al^{3+}$ is very difficult to detect because it does not form many water-insoluble solids. The usual procedure for identifying $Al^{3+}$ involves formation of $Al(OH)_3$, but this method is difficult, since $Al(OH)_3$ is a gray, translucent, hard-to-see solid that dissolves both in acid and in base. Aluminon is a dye that stains $Al(OH)_3$ to increase its visibility. It does not react with $Al^{3+}$ directly.
3. Use of $K_2CrO_4$ early in a separation scheme often leads to problems, since $CrO_4^{2-}$ is a powerful oxidizing agent. For example, $CrO_4^{2-}$ reacts with $S^{2-}$ to produce sulfur and green $Cr^{3+}$.
4. Although you should try hard to find a scheme in which the ions are completely separated, it sometimes may be necessary to test a pair of ions in the same solution. To do this successfully, you must be sure that the ions do not interfere with each other. Confirm this using known samples.

5. The secret to success in this experiment is to do the known carefully. Because the solutions encountered in this experiment are complicated, it is difficult to predict behavior using $K_{sp}$ calculations alone. To see what happens under a particular set of circumstances, try a known.
6. When reactions occur in a mixture of ions that are not expected on the basis of the tests on the individual ions, the problem most likely involves one of the following:
   (a) Wrong pH. Salts with basic anions are soluble in acidic solution.
   (b) Formation of a complex ion. The concentration of free metal ion can be too low to allow precipitation.

Complete the following exercises after you have carried out the tests on the individual cations.

## EXERCISES

1. Giving all reagents, reaction conditions, and equations, indicate how you would separate the following mixtures of cations in aqueous solution:
   (a) $Zn^{2+}$, $Al^{3+}$
   (b) $Hg^{2+}$, $Fe^{3+}$
   (c) $Ba^{2+}$, $Pb^{2+}$ (give two methods)
   (d) $Fe^{3+}$, $Bi^{3+}$, $Ag^{+}$
   (e) $Ni^{2+}$, $Cu^{2+}$
2. Give the net ionic equations for the following reactions:
   (a) Barium chloride with potassium chromate.
   (b) Aqueous ammonia with mercury(II) nitrate.
   (c) The oxidation of manganese(II) ion with sodium bismuthate in acid solution.
   (d) Sodium phosphate with nickel(II) chloride.
3. Consider the following separation scheme:

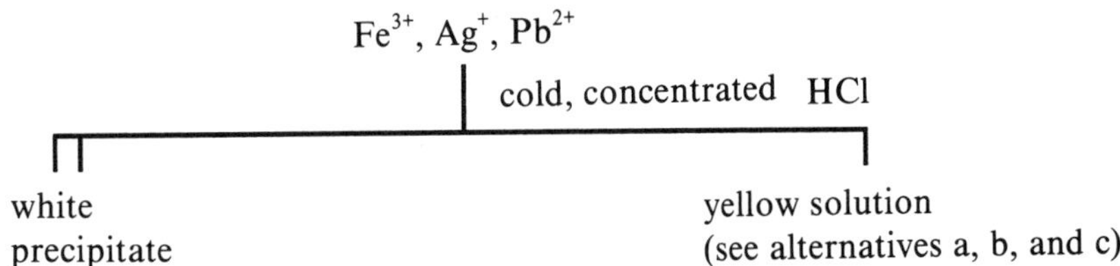


   (a) Saturate the yellow solution with $H_2S$. The filtrate develops a fine yellow or tan precipitate. Explain.
   (b) Add NaOH to the yellow solution. A red-orange precipitate develops. Explain.
   (c) Raise the pH of the yellow solution to about 1 or 2, add 1–2 mL of thioacetamide, and heat. A tan precipitate develops, and the solution turns black on addition of NaOH. Explain.
4. When acid is added to a carbonate salt or a solution containing carbonate, a gas is evolved. What is the gas? Write an equation to account for the formation of this gas.
5. Although the very insoluble metal sulfide solids do not dissolve in HCl(*aq*), most do dissolve in $HNO_3$(*aq*). Explain the difference in the behavior of sulfide solids toward these two acids.

**Useful References**

C.H. Sorum. *Introduction to Semimicro Qualitative Analysis.* 6th ed. Englewood Cliffs, N.J.: Prentice-Hall, 1983.

D.C. Layde and D.H. Busch. *Introduction to Qualitative Analysis.* 2d.ed. Newton, Mass.: Allyn and Bacon, 1968.

Therald Moeller. *Qualitative Analysis.* New York: McGraw-Hill, 1958.

# Complete Solutions to Tests

## Chapter 1

### Test 1.1

**A.** See discussion in the text.

**B.**

1. $K = \dfrac{[SO_3]^2}{[SO_2]^2[O_2]}$

2. $K = \dfrac{[O_3]^2}{[O_2]^3}$

3. $K = \dfrac{[NO_2]^4[H_2O]^6}{[NH_3]^4[O_2]^7}$

4. $K = \dfrac{[H_2O]^3}{[H_2]^3}$

Note: Water is included because it is in the gas phase.

**C.**

1. $K = \dfrac{[D]^2[C]^3}{[B]^2[A]}$

2. $K = \dfrac{(0.50 \text{ mol/L})^2(1.0 \text{ mol/L})^3}{(0.20 \text{ mol/L})^2(0.35 \text{ mol/L})} = 18 \text{ (mol/L)}^2$

$\Delta n = 5 - 3 = 2$

$K_p = K \cdot (RT)^{\Delta n}$

$T = 298 \text{ K}, R = 0.0821 \text{ L} \cdot \text{atm/K} \cdot \text{mol}$

$$K_p = \left[18\left(\frac{\text{mol}}{\text{L}}\right)^2\right]\left[\left(0.0821\ \frac{\text{L} \cdot \text{atm}}{\text{K} \cdot \text{mol}}\right)(298 \text{ K})\right]^2$$

$$= 1.1 \times 10^4 \text{ atm}^2$$

### Test 1.2

Some $NH_3$ decomposes to form 0.399 mol of $N_2$ and an unknown amount of $H_2$. Using the stoichiometry dictated by the balanced equation, the amounts of $NH_3$ and $H_2$ present at equilibrium can be calculated:

$$0.399 \text{ mol } N_2 \times \frac{2 \text{ mol } NH_3}{1 \text{ mol } N_2} = 0.798 \text{ mol of } NH_3 \text{ consumed to produce } 0.399 \text{ mol of } N_2$$

$$NH_3 \text{ remaining} = 1.000 - 0.798 = 0.202 \text{ mol}$$

At equilibrium

$$[NH_3] = [NH_3]_0 \text{ plus change to reach equilibrium}$$

$$= \frac{1.000 \text{ mol}}{1.000 \text{ L}} - \frac{0.798 \text{ mol}}{1.000 \text{ L}} = 0.202 \text{ mol/L}$$

The amount of $H_2$ produced is given by

$$0.399 \text{ mol } N_2 \times \frac{3 \text{ mol } H_2}{1 \text{ mol } N_2} = 1.197 \text{ mol } H_2$$

Since no $H_2$ was originally present in the flask,

$$[H_2] = [H_2]_0 \text{ plus change to reach equilibrium}$$
$$= 0 + \frac{1.197 \text{ mol}}{1.000 \text{ L}} = 1.197 \text{ mol/L}$$

**Test 1.3**

A. $K = \dfrac{P_{NO_2}{}^2}{P_{N_2O_4}}$

B. $P_{N_2O_4}^{\text{initial}} = 2.0$ atm; $P_{N_2O_4}^{\text{equilibrium}} = 1.8$ atm;
$P_{N_2O_4}^{\text{initial}} - P_{N_2O_4}^{\text{equilibrium}} = 0.2$ atm

Thus, 0.2 atm of $N_2O_4(g)$ was consumed as the system proceeded to equilibrium.

C. Since $N_2O_4(g) \longrightarrow 2NO_2(g)$, 0.2 atm $N_2O_4(g)$ produces 0.4 atm $NO_2(g)$.

D. $K = \dfrac{P_{NO_2}{}^2}{P_{N_2O_4}} = \dfrac{(0.4)^2}{(1.8)} = 9 \times 10^{-2}$ atm

**Test 1.4**

For each experiment (I, II, and III), the value of $K$ is $6 \times 10^{-2}$. That is,

$$\frac{[NH_3]^2}{[N_2][H_2]^3} = 6 \times 10^{-2}$$

for each case.

**Test 1.5**

A. 1. The initial concentrations are:

$$[SO_2]_0 = 2.00 \text{ mol/L}$$
$$[O_2]_0 = 1.50 \text{ mol/L}$$
$$[SO_3]_0 = 3.00 \text{ mol/L}$$

At equilibrium, $[SO_3] = 3.50$ mol/L. The increase in $SO_3$ concentration is 3.50 mol/L − 3.00 mol/L, or 0.50 mol/L of $SO_3$. Now calculate the corresponding changes in $[O_2]$ and $[SO_2]$.

$$0.50 \text{ mol } SO_3 \times \frac{2 \text{ mol } SO_2}{2 \text{ mol } SO_3} = 0.50 \text{ mol of } SO_2 \text{ formed to reach equilibrium}$$

The change in $[SO_2]$ is negative 0.50 mol/1.00 L, since some $SO_2$ must be consumed.

$$0.50 \text{ mol } SO_3 \times \frac{1 \text{ mol } O_2}{2 \text{ mol } SO_3} = 0.25 \text{ mol of } O_2 \text{ formed to reach equilibrium}$$

The change in $[O_2]$ is negative 0.25 mol/1.00 L, since some $O_2$ must be consumed. Now compute the equilibrium concentrations.

$$[SO_2] = [SO_2]_0 \text{ plus change}$$
$$= \frac{2.00 \text{ mol}}{1.00 \text{ L}} - \frac{0.50 \text{ mol}}{1.00 \text{ L}} = 1.50 \text{ mol/L}$$
$$[O_2]_0 = [O_2]_0 \text{ plus change}$$
$$= \frac{1.50 \text{ mol}}{1.00 \text{ L}} - \frac{0.25 \text{ mol}}{1.00 \text{ L}} = 1.25 \text{ mol/L}$$

2. $K = \dfrac{[SO_3]^2}{[SO_2]^2[O_2]} = \dfrac{(3.50\ \text{mol/L})^2}{(1.50\ \text{mol/L})^2(1.25\ \text{mol/L})} = 4.36\ \text{L/mol}$

3. $K_p = K(RT)^{\Delta n}$

$\Delta n = 2 - (2 + 1) = -1$

$K = 4.36\ \text{L/mol}$

$T = 600°\text{C} = 600 + 273 = 873\ \text{K}$

$R = 0.08206\ \text{L} \cdot \text{atm/K} \cdot \text{mol}$

$$K_p = (4.36)(RT)^{-1} = \frac{4.36}{RT} = \frac{4.36\ \text{L/mol}}{\left(0.08206\ \dfrac{\text{L} \cdot \text{atm}}{\text{K} \cdot \text{mol}}\right)(873\ \text{K})}$$

$= 6.08 \times 10^{-2}\ \text{atm}^{-1}$

**B.** 1. The initial concentrations are

$$[SO_2]_0 = 0.500\ \text{mol/L}$$
$$[SO_3]_0 = 0.350\ \text{mol/L}$$
$$[O_2]_0 = 0$$

The concentration of $O_2$ at equilibrium is

$$[O_2] = 0.045\ \text{mol/L}$$

Thus, to reach equilibrium, 0.045 mol/L of $O_2$ has been formed. This means that $SO_3$ has been consumed to form $SO_2$ and $O_2$. Compute the changes needed to reach equilibrium.

Change in $[O_2]$: 0.045 mol/L
Change in $[SO_3]$:

$$0.045\ \text{mol/L}\ O_2 \times \frac{2\ \text{mol}\ SO_3}{1\ \text{mol}\ O_2} = 0.090\ \text{mol/L of}\ SO_3$$ *consumed* to reach equilibrium

Change in $[SO_2]$:

$$0.045\ \text{mol/L}\ O_2 \times \frac{2\ \text{mol}\ SO_2}{1\ \text{mol}\ O_2} = 0.090\ \text{mol/L of}\ SO_2$$ *formed* to reach equilibrium

Now calculate the equilibrium concentrations.

$$[O_2] = [O_2]_0 \text{ plus change to reach equilibrium}$$
$$= 0 + 0.045\ \text{mol/L} = 0.045\ \text{mol/L}$$
$$[SO_3] = [SO_3]_0 \text{ plus change to reach equilibrium}$$
$$= 0.350\ \text{mol/L} - 0.090\ \text{mol/L} = 0.260\ \text{mol/L}$$
$$[SO_2] = [SO_2]_0 \text{ plus change to reach equilibrium}$$
$$= 0.500\ \text{mol/L} + 0.090\ \text{mol/L} = 0.590\ \text{mol/L}$$

2. $K = \dfrac{[SO_3]^2}{[SO_2]^2[O_2]} = \dfrac{(0.260\ \text{mol/L})^2}{(0.590\ \text{mol/L})^2(0.045\ \text{mol/L})} = 4.32\ \text{L/mol}$

Note that $K$ calculated here has the same value (within the error caused by round-off) as calculated in part A. This again demonstrates the constancy of $K$ at a given temperature. The experiments in parts A and B produce different equilibrium positions, because the initial concentrations were different, but give the same ratio defined by the equilibrium expression.

**Test 1.6**

The initial concentrations are

$$[NOCl]_0 = \frac{1.0 \times 10^{-1}\ \text{mol}}{2.0\ \text{L}} = 5.0 \times 10^{-2}\ \text{mol/L}$$

$$[NO]_0 = \frac{1.0 \times 10^{-3}\ \text{mol}}{2.0\ \text{L}} = 5.0 \times 10^{-4}\ \text{mol/L}$$

$$[Cl_2]_0 = \frac{1.0 \times 10^{-4}\ \text{mol}}{2.0\ \text{L}} = 5.0 \times 10^{-5}\ \text{mol/L}$$

$$Q = \frac{[NO]_0^2[Cl_2]_0}{[NOCl]_0^2} = \frac{(5.0 \times 10^{-4})^2(5.0 \times 10^{-5})}{(5.0 \times 10^{-2})^2}$$

$$= 5.0 \times 10^{-9}\ \text{mol/L}$$

$K$ is $1.55 \times 10^{-5}$ mol/L. Thus $Q$ is less than $K$, and the system will proceed to the right to reach equilibrium.

**Test 1.7**

The equilibrium expression is

$$K = 5.0 = \frac{[CO_2][H_2]}{[CO][H_2O]}$$

The initial concentrations are

$$[CO]_0 = \frac{3.00\ \text{mol}}{1.00\ \text{L}} = 3.00\ \text{mol/L}$$

$$[H_2O]_0 = \frac{0.50\ \text{mol}}{1.00\ \text{L}} = 3.00\ \text{mol/L}$$

$$[CO_2]_0 = 0$$

$$[H_2]_0 = 0$$

Let $x$ be the number of mol/L of $CO_2$ formed as the system reaches equilibrium. This means that

$x$ mol/L of $H_2$ are formed
$x$ mol/L of $H_2O$ are consumed
$x$ mol/L of CO are consumed

The equilibrium concentrations are

$$[CO_2] = [CO_2]_0 + x = 0 + x\ \text{mol/L}$$
$$[H_2] = [H_2]_0 + x = 0 + x\ \text{mol/L}$$
$$[CO] = [CO]_0 - x = 3.00\ \text{mol/L} - x\ \text{mol/L}$$
$$[H_2O] = [H_2O]_0 - x = 3.00\ \text{mol/L} - x\ \text{mol/L}$$

Now substitute these equilibrium concentrations into the equilibrium constant expression:

$$K = 5.10 = \frac{[CO_2][H_2]}{[CO][H_2O]} = \frac{(x)(x)}{(3.00 - x)(3.00 - x)} = \frac{x^2}{(3.00 - x)^2}$$

Taking the square root of both sides:

$$\frac{x}{(3.00 - x)} = \sqrt{5.10} = 2.26$$

$$x = 2.26(3.00 - x) = 6.78 - 2.26x$$

$$3.26x = 6.78$$

$$x = \frac{6.78}{3.26} = 2.08 \text{ mol/L}$$

Now compute the equilibrium concentrations:

$$[CO_2] = x = 2.08 \text{ mol/L}$$
$$[H_2] = x = 2.08 \text{ mol/L}$$
$$[CO] = 3.00 - x = 0.92 \text{ mol/L}$$
$$[H_2O] = 3.00 - x = 0.92 \text{ mol/L}$$

**Test 1.8**

The equilibrium expression is

$$K = 60.0 = \frac{[HI]^2}{[H_2][I_2]}$$

The initial concentrations are

$$[H_2]_0 = \frac{1.50 \text{ mol}}{1.00 \text{ L}} = 1.50 \text{ mol/L}$$

$$[I_2]_0 = \frac{2.50 \text{ mol}}{1.00 \text{ L}} = 2.50 \text{ mol/L}$$

$$[HI]_0 = 0$$

Assume that $x$ mol/L of $H_2$ are consumed to reach equilibrium. Using the balanced equation:

$$x \text{ mol/L } H_2 + x \text{ mol/L } I_2 \longrightarrow 2x \text{ mol/L HI}$$

The equilibrium concentrations are

$$[H_2] = [H_2]_0 + \text{change} = 1.50 \text{ mol/L} - x \text{ mol/L}$$
$$[I_2] = [I_2]_0 + \text{change} = 2.50 \text{ mol/L} - x \text{ mol/L}$$
$$[HI] = [HI]_0 + \text{change} = 0 + 2x \text{ mol/L}$$

Now substitute into the equilibrium constant expression:

$$K = 60.0 = \frac{(2x)^2}{(1.50 - x)(2.50 - x)} = \frac{4x^2}{3.75 - 4.00x + x^2}$$

Multiply this out and collect terms:

$$225 - 240x + 60.0x^2 = 4x^2$$
$$56.0x^2 - 240x + 225 = 0$$

This is a quadratic equation, where $a = 56.0$, $b = -240$, $c = 225$. Substituting into the quadratic formula,

$$x = \frac{-b \pm \sqrt{b^2 - 4ac}}{2a} = \frac{-(-240) \pm \sqrt{(-240)^2 - 4(56.0)(225)}}{2(56.0)}$$

The two roots are $x = 2.90$ and $x = 1.38$. Because $[H_2] = 1.50 - x$, $x = 2.90$ cannot be correct, since the $[H_2]$ cannot be negative. Thus $x = 1.38$ mol/L is the physically correct root. Now calculate the equilibrium concentrations:

$$[H_2] = 1.50 - x = 1.50 \text{ mol/L} - 1.38 \text{ mol/L} = 0.12 \text{ mol/L}$$
$$[I_2] = 2.50 - x = 2.50 \text{ mol/L} - 1.38 \text{ mol/L} = 1.12 \text{ mol/L}$$
$$[HI] = 2x = 2(1.38 \text{ mol/L}) = 2.76 \text{ mol/L}$$

**Test 1.9**

A. $[PCl_5]_0 = \frac{0.200 \text{ mol}}{2.00 \text{ L}} = 0.100 \text{ mol/L}$

$[PCl_3]_0 = \frac{0.500 \text{ mol}}{2.00 \text{ L}} = 0.250 \text{ mol/L}$

$[Cl_2]_0 = \frac{0.300 \text{ mol}}{2.00 \text{ L}} = 0.150 \text{ mol/L}$

B. $Q = \frac{[PCl_3][Cl_2]}{[PCl_5]} = \frac{(0.250)(0.150)}{(0.100)} = 0.375$

C. $Q$ is greater than $K$ (which equals $5.0 \times 10^{-2}$). Thus the system will adjust to the left to come to equilibrium.

D. Let $x$ = mol/L of $PCl_5(g)$ formed to come to equilibrium.

E. $[PCl_5] = 0.100 + x$
$[PCl_3] = 0.250 - x$
$[Cl_2] = 0.150 - x$

F. $K = \frac{[PCl_3][Cl_2]}{[PCl_5]} = \frac{(0.250 - x)(0.150 - x)}{(0.100 + x)} = 5.0 \times 10^{-2}$

This produces the quadratic equation $x^2 - 0.450x + 0.0325 = 0$.
Using the quadratic formula, $x = 0.090$ mol/L.

G. $[PCl_5] = 0.100 + 0.090 = 0.190$ mol/L
$[PCl_3] = 0.250 - 0.090 = 0.160$ mol/L
$[Cl_2] = 0.150 - 0.090 = 0.060$ mol/L

**Test 1.10**

A. Shifts to the right.

B. Shifts to the right.

C. Shifts to the left (the reactants have smaller volume than the products).

D. No effect.

E. Shifts to the left. The reaction is exothermic:

$$2NO_2(g) \rightleftarrows N_2(g) + 2O_2(g) + \text{heat}$$

Added heat will drive the equilibrium to the left. (In this case, $K$ will decrease with increasing temperature.)

## Chapter 2

**Test 2.1**

A. See discussion in the chapter.

B. 1. $HCN \rightleftarrows H^+ + CN^-$
2. $C_6H_5NH_3^+ \rightleftarrows C_6H_5NH_2 + H^+$
3. $Al(OH_2)_6^{3+} \rightleftarrows Al(OH)(OH_2)_5^{2+} + H^+$

**C.** This reaction is the reverse of the $K_a$ reaction for HCN:

$$K = \frac{[HCN]}{[H^+][CN^-]} = \frac{1}{K_a} = \frac{1}{6.2 \times 10^{-10}} = 1.6 \times 10^9$$

**D.** 1. $OCl^-$ is the strongest base. This is clear because $K_a$ for HOCl is the smallest of all those listed.

2. Base strength (use the $K_a$ values for the acids):

$$OCl^- > NO_2^- > F^- > IO_3^-$$

strongest base (smallest $K_a$ for HA) — weakest base (largest $K_a$ for HA)

3. Acid strength:

$$HOCl < HNO_2 < HF < HIO_3$$

weakest (smallest $K_a$) — strongest (largest $K_a$)

### Test 2.2

**A.** $[OH^-] = 2.0 \times 10^{-4}\ M$

$[H^+][OH^-] = 1.0 \times 10^{-14}$

$[H^+](2.0 \times 10^{-4}) = 1.0 \times 10^{-14}$

$$[H^+] = \frac{1.0 \times 10^{-14}}{2.0 \times 10^{-4}} = 5.0 \times 10^{-11}\ M$$

$pH = -\log[H^+] = -\log(5.0 \times 10^{-11}) = 10.30$

**B.** $[H^+] = 5.0 \times 10^{-3}\ M$

$[H^+][OH^-] = 1.0 \times 10^{-14} = (5.0 \times 10^{-3})[OH^-]$

$$[OH^-] = \frac{1.0 \times 10^{-14}}{5.0 \times 10^{-3}} = 2.0 \times 10^{-12}\ M$$

$pOH = -\log[OH^-] = -\log(2.0 \times 10^{-12}) = 11.70$

### Test 2.3

**A.** 1. $HNO_3$, $H_2O$. The solution really contains $H^+$, $NO_3^-$, $H_2O$ as major species.
↑ strong acid

2. 0.1 $M$ $H^+$ from $HNO_3$ is the major source. $H_2O$ is the minor source.

3. Assume that the contribution of water to $[H^+]$ can be neglected.

4. $[H^+] = 0.10\ M = 1.0 \times 10^{-1}\ M$

$pH = -\log[H^+] = -\log(1.0 \times 10^{-1}) = 1.00$

**B.** 1. HCl is a strong acid: $H^+$, $Cl^-$, $H_2O$

2. $H^+$ from acid and dissociation of $H_2O$ are the sources of $H^+$.

3. In this case,

$$[H^+]_{HCl} = 1.0 \times 10^{-10}\ M$$

This is very small compared to the amount from dissociation of water ($1.0 \times 10^{-7}\ M$). $H_2O$ is the dominant source of $H^+$.

4. $[H^+] = 1.0 \times 10^{-7}$

$pH = 7.00$

Here, the strong acid is so dilute that the solution is still netural within the uncertainties of measurement.

## Test 2.4

**A.** 1. HOCl, $H_2O$

2. HOCl is a stronger acid than $H_2O$ ($K_a$ for HOCl is much larger than $K_w$) and will dominate in the production of $H^+$.

3. $HOCl \rightleftarrows H^+ + OCl^-$ $\quad K_a = \dfrac{[H^+][OCl^-]}{[HOCl]} = 3.5 \times 10^{-8}$

4. No dissociation has occurred.
$[HOCl]_0 = 0.100\ M$
$[OCl^-]_0 = 0$
$[H^+]_0 \approx 0.$; neglect contribution from $H_2O$.

5. $x =$ mol/L of HOCl that will dissociate as the system comes to equilibrium.

6. $[HOCl] = 0.100 - x$
$[OCl^-] = x$
$[H^+] = x$

7, 8. $3.5 \times 10^{-8} = \dfrac{[H^+][OCl^-]}{[HOCl]} = \dfrac{(x)(x)}{(0.100 - x)} \approx \dfrac{x^2}{0.100}$
$x^2 \approx (3.5 \times 10^{-8})(0.100) = 3.5 \times 10^{-9}$
$x \approx 5.9 \times 10^{-5}$

9. $\dfrac{5.9 \times 10^{-5}}{0.100} \times 100 = 5.9 \times 10^{-2}\%$
$0.100 \gg 5.9 \times 10^{-5}$
$0.100 - 5.9 \times 10^{-5} = 0.100$ (assumption is valid)

10. $[H^+] = x = 5.9 \times 10^{-5}$
$pH = -\log(5.9 \times 10^{-5}) = 4.23$

11. $[OCl^-] = [H^+] = 5.9 \times 10^{-5}$

12. $\%\text{ dissociation} = \dfrac{[OCl^-]}{[HOCl]_0} \times 100 = \dfrac{5.9 \times 10^{-5}}{(0.100)} \times 100 = 5.9 \times 10^{-2}\%$

13. $[H^+][OH^-] = K_w = 1.0 \times 10^{-14}$
$(5.9 \times 10^{-5})[OH^-] = 1.0 \times 10^{-14}$
$[OH^-] = 1.7 \times 10^{-10}\ M$

**B.** In solution, we have $H_3BO_3$ (boric acid) and $H_2O$. Boric acid is a very weak acid ($K_a = 5.8 \times 10^{-10}$) but is stronger than $H_2O$ ($K_w = 10^{-14}$). Thus the boric acid dissociation will determine the $[H^+]$:

$$H_3BO_3 \rightleftarrows H^+ + H_2BO_3^- \qquad K_a = \frac{[H^+][H_2BO_3^-]}{[H_3BO_3]} = 5.8 \times 10^{-10}$$

| *Initial Concentrations (before any dissociation)* | | *Equilibrium Concentrations* |
|---|---|---|
| $[H_3BO_3]_0 = 0.50\ M$ | Let $x$ mol/L of $H_3BO_3$ dissociate → | $[H_3BO_3] = 0.50 - x$ |
| $[H_2BO_3^-]_0 = 0$ | | $[H_2BO_3^-] = x$ |
| $[H^+]_0 \approx 0$ | | $[H^+] = x$ |

$$5.8 \times 10^{-10} = K_a = \frac{[H^+][H_2BO_3^-]}{[H_3BO_3]} = \frac{(x)(x)}{0.50 - x} \approx \frac{x^2}{0.5}$$

$$x^2 \approx 2.9 \times 10^{-10}$$

$$x \approx 1.7 \times 10^{-5}$$

Note that $0.50 \gg 1.7 \times 10^{-5}$, so $[H^+] = x = 1.7 \times 10^{-5}\ M$.
$pH = -\log(1.7 \times 10^{-5}) = 4.77$

**Test 2.5**

The solution contains HA, $Na^+$, $A^-$, $H_2O$. The equilibrium of interest is:

$HA \rightleftarrows H^+ + A^-$

$[H^+] = 10^{-pH} = 1.0 \times 10^{-4}$

$$K_a = \frac{[H^+][A^-]}{[HA]} = \frac{(1.0 \times 10^{-4})(0.025)}{(0.050)} = 5.0 \times 10^{-5}$$

**Test 2.6**

**A.** The solution contains the major species $K^+$, $OH^-$, and $H_2O$.

**B.** The $OH^-$ from the dissolved KOH.

**C.** Assume that $H_2O$ produces a neglible amount of $OH^-$, so $[OH^-] = 0.10\ M$.

**D.** $[H^+][OH^-] = K_w = 1.00 \times 10^{-14}$

$[H^+](0.10) = 1.00 \times 10^{-14}$

$$[H^+] = \frac{(1.00 \times 10^{-14})}{(0.10)} = 1.0 \times 10^{-13}\ M$$

**E.** $pH = -\log[H^+] = -\log(1.0 \times 10^{-13}) = 13.00$

**Test 2.7**

**A.** A base reacting with $H_2O$ to produce $OH^-$ and the conjugate acid.

**B.** See the discussion in the chapter.

**C.** $K_b \neq \dfrac{1}{K_a}$

$B + H_2O \rightleftarrows BH^+ + OH^- \qquad K_b = \dfrac{[BH^+][OH^-]}{[B]}$

$BH^+ \rightleftarrows B + H^+ \qquad K_a = \dfrac{[B][H^+]}{[BH^+]}$

$B + H^+ \rightleftarrows BH^+ \qquad K = \dfrac{1}{K_a} = \dfrac{[BH^+]}{[B][H^+]}$

$K_b$ and $\dfrac{1}{K_a}$ refer to entirely different reactions.

**D.**

1. $CH_3NH_2$, $H_2O$
2. $CH_3NH_2 + H_2O \rightleftarrows CH_3NH_3^+ + OH^-$
3. $K_b$ for $CH_3NH_2$ ($4.38 \times 10^{-4}$) is much greater than $K_w$. $CH_3NH_2$ will dominate.
4. $K_b = \dfrac{[CH_3NH_3^+][OH^-]}{[CH_3NH_2]}$
5. $[CH_3NH_2]_0 = 1.0\ M$
   $[CH_3NH_3^+]_0 = 0$
   $[OH^-] \approx 0$ (neglect contribution of $H_2O$)
6. $x$ = mol/L of $CH_3NH_2$ that react with $H_2O$ to come to equilibrium
7. $[CH_3NH_2] = 1.0 - x$
   $[CH_3NH_3^+] = x$
   $[OH^-] = x$
8. $4.38 \times 10^{-4} = K_b = \dfrac{[CH_3NH_3^+][OH^-]}{[CH_3NH_2]} = \dfrac{(x)(x)}{(1.0 - x)} \approx \dfrac{x^2}{1.0}$
   $x^2 \approx 4.38 \times 10^{-4}$
   $x \approx 2.1 \times 10^{-2}$

$$\frac{x}{[CH_3NH_2]_0} \times 100 = \frac{2.1 \times 10^{-2}}{1.0} \times 100 = 2.1\%$$

The assumption that $1.0 - x \approx 1.0$ is acceptable.

9. $[OH^-] = x = 2.1 \times 10^{-2}\ M$
   $[H^+][OH^-] = 1.0 \times 10^{-14}$
   $[H^+](2.1 \times 10^{-2}) = 1.0 \times 10^{-14}$
   $$[H^+] = \frac{1.0 \times 10^{-14}}{2.1 \times 10^{-2}} = 4.8 \times 10^{-13}\ M$$
   pH = 12.32

**E.** In solution, the major species are $C_5H_5N$ and $H_2O$. $C_5H_5N$ is a base ($K_b = 1.4 \times 10^{-9}$), which will dominate $H_2O$ in the production of $OH^-$:

$$C_5H_5N + H_2O \rightleftarrows C_5H_5NH^+ + OH^-$$

$$K_b = \frac{[C_5H_5NH^+][OH^-]}{[C_5H_5N]} = 1.4 \times 10^{-9}$$

| *Initial Concentrations* | | *Equilibrium Concentrations* |
|---|---|---|
| $[C_5H_5N]_0 = 0.10\ M$ | Let $x$ mol/L $C_5H_5N$ react with $H_2O$ to reach equilibrium → | $[C_5H_5N] = 0.10 - x$ |
| $[C_5H_5NH^+]_0 = 0$ | | $[C_5H_5NH^+] = x$ |
| $[OH^-]_0 \approx 0$ | | $[OH^-] = x$ |

$$1.4 \times 10^{-9} = K_b = \frac{[C_5H_5NH^+][OH^-]}{[C_5H_5N]} = \frac{(x)(x)}{(0.10 - x)} \approx \frac{x^2}{0.10}$$

$x^2 \approx (1.4 \times 10^{-9})(0.10) = 1.4 \times 10^{-10}$
$x \approx 1.2 \times 10^{-5}$

Assumption that $0.1 - x \approx 0.1$ is valid.
$[OH^-] = x = 1.2 \times 10^{-5}$
$[H^+][OH^-] = K_w = 1.0 \times 10^{-14}$

$$[H^+] = \frac{1.0 \times 10^{-14}}{[OH^-]} = \frac{1.0 \times 10^{-14}}{1.2 \times 10^{-5}} = 8.4 \times 10^{-10}$$

pH = 9.08

## Test 2.8

**A.** The solution contains HOAc, $OAc^-$, $Na^+$, $H_2O$. The acetic acid dissociation equilibrium will control the $[H^+]$:

$$HOAc \rightleftarrows H^+ + OAc^- \qquad K_a = \frac{[H^+][OAc^-]}{[HOAc]} = 1.8 \times 10^{-5}$$

| *Initial Concentrations* | | *Equilibrium Concentrations* |
|---|---|---|
| $[HOAc]_0 = 0.20\ M$ | Let $x$ mol/L HOAc dissociate → | $[HOAc] = 0.20 - x$ |
| $[OAc^-]_0 = 0.10\ M$ | | $[OAc^-] = 0.10 + x$ |
| $[H^+]_0 \approx 0$ | | $[H^+] = x$ |

$$1.8 \times 10^{-5} = K_a = \frac{[H^+][OAc^-]}{[HOAc]} = \frac{(x)(0.10 + x)}{0.20 - x} \approx \frac{(x)(0.10)}{0.20}$$

$x \approx 3.6 \times 10^{-5}$

Assumptions are valid. $[H^+] = 3.6 \times 10^{-5}$; pH = 4.44.

**B.** Before any reaction, the solution contains the major species $H^+$, $Cl^-$, $Na^+$, $OAc^-$, HOAc, $H_2O$. The reaction that will occur is

| | $H^+$ | + | $OAc^-$ | $\longrightarrow$ | HOAc |
|---|---|---|---|---|---|
| Before reaction | 0.02 mol | | 0.10 mol | | 0.20 mol |
| After reaction | 0 | | $0.10 - 0.02 = 0.080$ mol | | $0.20 + 0.02 = 0.22$ mol |

After the reaction, the solution contains HOAc, $OAc^-$, $Na^+$, $Cl^-$, $H_2O$. The equilibrium of interest is

$$HOAc \rightleftarrows H^+ + OAc^- \qquad K_a = \frac{[H^+][OAc^-]}{[HOAc]} = 1.8 \times 10^{-5}$$

*Initial Concentrations*

$$[HOAc]_0 = \frac{0.22 \text{ mol}}{1.0 \text{ L}} = 0.22\ M$$

$$[OAc^-]_0 = \frac{0.080 \text{ mol}}{1.0 \text{ L}} = 0.080\ M$$

$$[H^+]_0 \approx 0$$

Let $x$ mol/L HOAc dissociate $\longrightarrow$

*Equilibrium Concentrations*

$$[HOAc] = 0.22 - x$$

$$[OAc^-] = 0.08 + x$$

$$[H^+] = x$$

$$1.8 \times 10^{-5} = K_a = \frac{[H^+][OAc^-]}{[HOAc]} = \frac{(x)(0.080 + x)}{(0.22 - x)} \approx \frac{(x)(0.080)}{(0.22)}$$

$$x \approx \frac{(0.22)(1.8 \times 10^{-5})}{(0.080)} = 5.0 \times 10^{-5}$$

Assumptions are valid. $[H^+] = 4.9 \times 10^{-5}$; pH = 4.30. Thus, adding 0.02 mol of HCl($g$) to this solution changes the pH from 4.44 to 4.30. This is a relatively small change.

**C.** Before any reaction, the solution contains the major species $Na^+$, $OH^-$, $OAc^-$, HOAc, $H_2O$. The reaction that will occur is

| | $OH^-$ | + | HOAc | $\longrightarrow$ | $OAc^-$ | + $H_2O$ |
|---|---|---|---|---|---|---|
| Before reaction | 0.02 mol | | 0.20 mol | | 0.10 mol | |
| After reaction | 0 | | $0.20 - 0.02 = 0.18$ mol | | $0.10 + 0.02 = 0.12$ mol | |

After the reaction, the solution contains $Na^+$, $OAc^-$, HOAc, $H_2O$. The equilibrium of interest is

$$HOAc \rightleftarrows H^+ + OAc^- \qquad K_a = \frac{[H^+][OAc^-]}{[HOAc]} = 1.8 \times 10^{-5}$$

*Initial Concentrations*

$$[HOAc]_0 = 0.18\ M$$

$$[OAc^-]_0 = 0.12\ M$$

$$[H^+]_0 \approx 0$$

Let $x$ mol/L HOAc dissociate $\longrightarrow$

*Equilibrium Concentrations*

$$[HOAc] = 0.18 - x$$

$$[OAc^-] = 0.12 + x$$

$$[H^+] = x$$

$$1.8 \times 10^{-5} = K_a = \frac{[H^+][OAc^-]}{[HOAc]} = \frac{(x)(0.12 + x)}{0.18 - x} \approx \frac{(x)(0.12)}{0.18}$$

$$x \approx \frac{(1.8 \times 10^{-5})(0.18)}{(0.12)} = 2.7 \times 10^{-5}$$

Assumptions are valid. $[H^+] = 2.7 \times 10^{-5}$; pH = 4.57. Thus, adding 0.02 mol of NaOH to the solution changes the pH from 4.44 to 4.57.

**Test 2.9**

**A.** I. $K_a = 1.0 \times 10^{-5} = \dfrac{[H^+][A^-]}{[HA]}$ where $[A^-] = [HA] = 1.0\ M$

$[H^+] = 1.0 \times 10^{-5}$; pH = 5.00

II. $K_a = 1.0 \times 10^{-5} = \dfrac{[H^+][A^-]}{[HA]}$ where $[A^-] = [HA] = 0.10\ M$

$[H^+] = 1.0 \times 10^{-5}$; pH = 5.00

Both solutions have the same pH, since the $[A^-]/[HA]$ is the same in both.

**B.** I has the greater capacity.

**C.** In each case, the reaction is

$$HA + OH^- \longrightarrow H_2O + A^-$$

and in each case 0.05 mol of HA will be changed to 0.05 mol of $A^-$. Thus, the [HA] will decrease by 0.05 mol/L.

I. New pH:

$K_a = 1.0 \times 10^{-5} = \dfrac{[H^+][A^-]}{[HA]}$ where $[A^-] = 1.05\ M$, $[HA] = 0.95\ M$

$[H^+] = 9.0 \times 10^{-6}$; pH = 5.04

II. New pH:

$K_a = 1.0 \times 10^{-5} = \dfrac{[H^+][A^-]}{[HA]}$ where $[A^-] = 0.15\ M$, $[HA] = 0.05\ M$

$[H^+] = 3.3 \times 10^{-6}$; pH = 5.48

The original pH for both solutions was 5.00. Note that I resists a change in pH much better than II.

**Test 2.10**

**A.** 1. Since $[HA] \approx [A^-]$,

$$[H^+] = K_a \frac{[HA]}{[A^-]} \approx K_a = 1.0 \times 10^{-6}$$

pH = 6.0

2. Before the reaction, the solution contains the major species HA, $A^-$, $Na^+$, $OH^-$, $H_2O$. $OH^-$ will react with HA as follows:

| | $OH^-$ | + HA | $\longrightarrow$ $A^-$ | + $H_2O$ |
|---|---|---|---|---|
| Before reaction | 0.10 mol | 0.50 mol | 0.50 mol | |
| After reaction | 0 | 0.50 − 0.10 = 0.40 mol | 0.50 + 0.10 = 0.60 mol | |

After the reaction, the solution contains the major species HA, $A^-$, $Na^+$, $H_2O$, and the equilibrium of interest is

$$HA \rightleftarrows H^+ + A^- \qquad K_a = \frac{[H^+][A^-]}{[HA]} = 1.0 \times 10^{-6}$$

| *Initial Concentrations (before dissociation of HA)* | *Equilibrium Concentrations* |
|---|---|
| $[HA]_0 = \dfrac{0.40\ \text{mol}}{1.0\ \text{L}} = 0.40\ M$ | $[HA] = 0.40\ M - x$ |

$$[A^-]_0 = \frac{0.60 \text{ mol}}{1.0 \text{ L}} = 0.60\ M \quad \xrightarrow[\text{dissociate}]{\text{Let } x \text{ mol/L HA}} \quad [A^-] = 0.60\ M + x$$

$$[H^+]_0 \approx 0 \qquad [H^+] = x$$

$$1.0 \times 10^{-6} = K_a = \frac{[H^+][A^-]}{[HA]} = \frac{(x)(0.60 + x)}{(0.40 - x)} \approx \frac{(x)(0.60)}{(0.40)}$$

$$x = [H^+] = 6.7 \times 10^{-7}\ M$$

Assumption is valid. $pH = -\log[H^+] = -\log(6.7 \times 10^{-7}) = 6.17$

3. The reaction is

| | $H^+$ + | $A^-$ | $\longrightarrow$ HA |
|---|---|---|---|
| Before reaction | 0.10 mol | 0.50 mol | 0.50 mol |
| After reaction | 0 | 0.50 − 0.10 = 0.40 mol | 0.50 + 0.10 = 0.60 mol |

*Initial Concentrations (before dissociation of HA)* — *Equilibrium Concentrations*

$$[HA]_0 = \frac{0.60 \text{ mol}}{1.0 \text{ L}} = 0.60\ M \qquad [HA] = 0.60\ M - x$$

$$[A^-]_0 = \frac{0.40 \text{ mol}}{1.0 \text{ L}} = 0.40\ M \quad \xrightarrow[\text{dissociate}]{\text{Let } x \text{ mol/L HA}} \quad [A^-] = 0.40\ M + x$$

$$[H^+]_0 \approx 0 \qquad [H^+] = x$$

$$1.0 \times 10^{-6} = K_a = \frac{[H^+][A^-]}{[HA]} = \frac{(x)(0.40 + x)}{(0.60 - x)} \approx \frac{(x)(0.40)}{(0.60)}$$

$$x = [H^+] = 1.5 \times 10^{-6}\ M$$

$$pH = -\log[H^+] = -\log(1.5 \times 10^{-6}) = 5.82$$

**B.** 1. In each solution $[H^+] = K_a = 7.2 \times 10^{-4}$. Since in each case $[HF] = [F^-]$, pH = 3.14 for each.

2. Solution II has the largest capacity.

3. In each case, the reaction when $H^+$ is added is

$$H^+ + F^- \longrightarrow HF$$

and in each case the $[F^-]$ will decrease by 0.050 mol/L and the [HF] will increase by 0.050 mol/L.

I. New pH:

$$7.2 \times 10^{-4} = K_a = \frac{[H^+][F^-]}{[HF]} = \frac{[H^+](0.45)}{(0.55)}$$

$$[H^+] = \frac{(0.55)(7.2 \times 10^{-4})}{(0.45)} = 8.8 \times 10^{-4};\ pH = 3.06$$

II. New pH:

$$7.2 \times 10^{-4} = K_a = \frac{[H^+](4.95)}{(5.05)}$$

$$[H^+] = 7.3 \times 10^{-4};\ pH = 3.13$$

II resists a change in pH better than I, because it contains larger amounts of buffering materials.

**C.** $HNO_2 \rightleftarrows H^+ + NO_2^-$

$$K_a = 4.0 \times 10^{-4} = \frac{[H^+][NO_2^-]}{[HNO_2]}$$

$pH = 3.00$

$[H^+] = 10^{-pH} = 1.0 \times 10^{-3}\ M$

$[HNO_2] = 0.050\ M$

$$4.0 \times 10^{-4} = \frac{(1.0 \times 10^{-3})[NO_2^-]}{(0.050)}$$

$$[NO_2^-] = \frac{(0.050)(4.0 \times 10^{-4})}{1.0 \times 10^{-3}} = 2.0 \times 10^{-2}\ M$$

0.020 mol of $NaNO_2(s)$ must be added to each liter of the 0.050 $M$ solution to produce a buffer with pH = 3.00.

**D.** Before the reaction, the major species are $H^+$, $Cl^-$, $Na^+$, $OAc^-$, $H_2O$. The reaction which will take place is

| | $H^+$ | + | $OAc^-$ | ⟶ | HOAc |
|---|---|---|---|---|---|
| Before reaction | 50.0 mL × 0.050 $M$ = 2.5 mmol | | 100.0 mL × 0.10 $M$ = 10.0 mmol | | 0 |
| After reaction | 0 | | 10.0 − 2.5 = 7.5 mmol | | 2.5 mmol |

In solution after the reaction: HOAc, $OAc^-$, $Na^+$, $H_2O$. The equilibrium to use is

$$HOAc \rightleftarrows H^+ + OAc^-$$

| *Initial Concentrations* | | *Equilibrium Concentrations* |
|---|---|---|
| $[HOAc]_0 = \frac{2.5 \text{ mmol}}{(100.0 + 50.0)\text{mL}} = 1.7 \times 10^{-2}\ M$ | Let $x$ mol/L HOAc dissociate ⟶ | $[HOAc] = 1.7 \times 10^{-2} - x$ |
| $[OAc^-]_0 = \frac{7.5 \text{ mmol}}{(100.0 + 50.0)\text{mL}} = 5.0 \times 10^{-2}\ M$ | | $[OAc^-] = 5.0 \times 10^{-2} + x$ |
| $[H^+]_0 \approx 0$ | | $[H^+] = x$ |

Plugging into the $K_a$ expression gives

$$x = 6.1 \times 10^{-6};\ pH = 5.21$$

**E.** Before the reaction, the major species are: $NH_4^+$, $Cl^-$, $Na^+$, $OH^-$, $H_2O$. The reaction will be

| | $OH^-$ | + | $NH_4^+$ | ⟶ | $NH_3$ | + $H_2O$ |
|---|---|---|---|---|---|---|
| Before reaction | 0.10 mol | | 0.500 mol | | 0 | |
| After reaction | 0 | | 0.40 mol | | 0.10 mol | |

After the reaction, the solution contains the major species $NH_4^+$, $Cl^-$, $NH_3$, $Na^+$ $H_2O$. The equilibrium to use is

$$NH_3 + H_2O \rightleftarrows NH_4^+ + OH^-$$

$$K_b = 1.8 \times 10^{-5} = \frac{[NH_4^-][OH^-]}{[NH_3]}$$

| *Initial Concentrations (before dissociation of $NH_4^+$)* | | *Equilibrium Concentrations* |
|---|---|---|
| $[NH_4^+]_0 = 0.40\ M$ | | $[NH_4^+] = 0.40 + x$ |
| $[NH_3]_0 = 0.10\ M$ | Let $x$ mol/L $NH_3$ react with $H_2O$ → | $[NH_3] = 0.10 - x$ |
| $[OH^-]_0 \approx 0$ | | $[OH^-] = x$ |

$$1.8 \times 10^{-5} = K_b = \frac{[NH_4^+][OH^-]}{[NH_3]}$$

$$= \frac{(0.40 + x)(x)}{(0.10 - x)} \approx \frac{(0.40)(x)}{0.10}$$

$$x \approx 4.5 \times 10^{-6}$$

The approximations are valid.

$$[OH^-] = x = 4.5 \times 10^{-6}$$

$$[H^+] = \frac{K_w}{[OH^-]} = \frac{1.0 \times 10^{-14}}{4.5 \times 10^{-6}} = 2.2 \times 10^{-9}$$

$$pH = 8.65$$

**Test 2.11**

**A.** 1. In solution: $C_6H_5NH_3^+$, $Cl^-$, $H_2O$

2. $K_a \cdot K_b = K_w$

$$K_a = \frac{K_w}{K_b} = \frac{1.0 \times 10^{-14}}{3.8 \times 10^{-10}} = 2.6 \times 10^{-5}$$

3. $C_6H_5NH_3^+ \rightleftarrows C_6H_5NH_2 + H^+$

4. $K_a = \dfrac{[C_6H_5NH_2][H^+]}{[C_6H_5NH_3^+]} = 2.6 \times 10^{-5}$

5. $[C_6H_5NH_3^+]_0 = 0.10\ M$
$[C_6H_5NH_2]_0 = 0$
$[H^+]_0 \approx 0$

6. $x =$ mol/L of $C_6H_5NH_3^+$ that dissociate to reach equilibrium.

7. $[C_6H_5NH_3^+] = 0.10 - x$
$[C_6H_5NH_2] = x$
$[H^+] = x$

8. $2.6 \times 10^{-5} = K_a = \dfrac{[C_6H_5NH_2][H^+]}{[C_6H_5NH_3^+]} = \dfrac{(x)(x)}{(0.10 - x)} \approx \dfrac{x^2}{0.10}$

$x^2 \approx (2.6 \times 10^{-5})(0.1) = 2.6 \times 10^{-6}$
$x = 1.6 \times 10^{-3}\ M = [H^+]$
Assumption is valid.

9. pH = 2.79

**B.** In solution: $Cr(OH_2)_6^{3+}$, $Cl^-$, $H_2O$. $Cr(OH_2)_6^{3+}$ is the strongest acid in solution:

$$Cr(OH_2)_6^{3+} \rightleftarrows Cr(OH)(OH_2)_5^{2+} + H^+$$

$$K_a = \frac{[Cr(OH)(OH_2)_5^{2+}][H^+]}{[Cr(OH_2)_6^{3+}]} = 1.5 \times 10^{-4}$$

| *Initial Concentrations* | *Equilibrium Concentrations* |
|---|---|
| $[Cr(OH_2)_6^{3+}]_0 = 0.20\ M$ | $[Cr(OH_2)_6^{3+}]_0 = 0.20 - x$ |

$[Cr(OH)(OH_2)_5^{2+}]_0 = 0$ $\xrightarrow[\text{dissociate}]{\text{Let } x \text{ mol/L } Cr(OH_2)_6^{3+}}$ $[Cr(OH)(OH_2)_5^{2+}] = x$

$[H^+]_0 \approx 0$ $[H^+] = x$

$$1.5 \times 10^{-4} = K_a = \frac{(x)(x)}{0.20 - x} \approx \frac{x^2}{0.20}$$

$$x^2 = (0.20)(1.5 \times 10^{-4}) = 3.0 \times 10^{-5}$$

$$x \approx 5.5 \times 10^{-3} = [H^+]$$

$$pH = 2.26$$

**Test 2.12**

**A.** 1. $Na^+$, $Cl^-$, $H_2O$; neutral; neither $Na^+$ nor $Cl^-$ is a significant acid or base.

2. $NH_4^+$, $NO_3^-$, $H_2O$. $NO_3^-$ is a *very* weak base ($HNO_3$ is a strong acid). $NH_4^+$ is a weak acid:

$$NH_4^+ \rightleftarrows NH_3 + H^+$$

The solution will be acidic.

3. $Al^{3+}_{(aq)}$, $NO_3^-$, $H_2O$. $Al^{3+}_{(aq)}$ is $Al(OH_2)_6^{3+}$, which is a weak acid:

$$Al(OH_2)_6^{3+} \rightleftarrows Al(OH)(OH_2)_5^{2+} + H^+$$

The solution will be acidic.

4. $Na^+$, $CN^-$, $H_2O$. $CN^-$ is a base (HCN is a very weak acid, $K_a = 6.2 \times 10^{-10}$):

$$CN^- + H_2O \rightleftarrows HCN + OH^-$$

The solution will be basic.

5. $(CH_3)_3NH^+$, $Cl^-$, $H_2O$. $(CH_3)_3NH^+$ is a weak acid ($K_a = \frac{K_w}{K_b} = (1.0 \times 10^{-14})/(5.3 \times 10^{-5}) = 1.9 \times 10^{-10}$):

$$(CH_3)_3NH^+ \rightleftarrows (CH_3)_3N + H^+$$

The solution will be acidic.

**B.** The statement that $OAc^-$ is a "good base" is based on the reaction:

$$HOAc \rightleftarrows H^+ + OAc^-$$

where the equilibrium lies to the left ($K_a = 1.8 \times 10^{-5}$). This reaction is really

$$HOAc + H_2O \rightleftarrows H_3O^+ + OAc^-$$

where $OAc^-$ and $H_2O$ are competing for $H^+$, and $OAc^-$ wins. In the reaction

$$OAc^- + H_2O \rightleftarrows HOAc + OH^-$$

$OAc^-$ is competing with $OH^-$, which is a *very* strong base. $OAc^-$ is a strong base compared to $H_2O$, but a weak base compared to $OH^-$. Thus the order of base strength is

$$OH^- > OAc^- > H_2O$$

## Chapter 3

**Test 3.1**

**A.** $[OH^-] = 0.50\ M$
$[H^+] = 2.0 \times 10^{-14}$ (using $K_w$)
$pH = 13.70$

**B.** In solution before the reaction are $H^+$, $Cl^-$, $Na^+$, $OH^-$, $H_2O$. The reaction is

| | $H^+$ | + | $OH^-$ | $\longrightarrow$ $H_2O$ |
|---|---|---|---|---|
| Before reaction | 25.0 mL × 1.0 *M* = 25 mmol | | 100.0 mL × 0.50 *M* = 50 mmol | |
| After reaction | 0 | | 50 − 25 = 25 | |

In solution after the reaction are $Na^+$, $Cl^-$, $OH^-$, $H_2O$.

$$[OH^-] = \frac{\text{mmol } OH^-}{\text{mL of solution}} = \frac{25}{100.0 + 25.0} = 2.0 \times 10^{-1}$$

$$[H^+] = \frac{K_w}{[OH^-]} = \frac{1.00 \times 10^{-14}}{2.0 \times 10^{-1}} = 5.0 \times 10^{-14}$$

$$pH = 13.30$$

**C.** This is the equivalence point. 50.0 mL × 1.0 *M* HCl = 50 mmol of HCl have been added, which exactly react with the 50 mmol of NaOH originally present. After this reaction, the solution contains $Na^+$, $Cl^-$, $H_2O$. Since neither $Na^+$ nor $Cl^-$ will affect the pH, then pH = 7.00.

**D.** Excess HCl has now been added. 75.0 mL × 1.0 *M* HCl = 75 mmol HCl added:

| 75 mmol | − | 50 mmol | = | 25 mmol HCl |
|---|---|---|---|---|
| ↑ total added | | ↑ consumed by NaOH | | ↑ excess |

The solution contains $H^+$, $Cl^-$, $Na^+$, $H_2O$.

$$[H^+] = \frac{\text{mmol HCl in excess}}{\text{mL of solution}} = \frac{25}{175.0} = 1.4 \times 10^{-1}\ M$$

$$pH = 0.85$$

### Test 3.2

**C.** 1. HOAc, $Na^+$, $OH^-$, $H_2O$

2. $OH^- + HOAc \longrightarrow H_2O + OAc^-$

3.

| | $OH^-$ | + | HOAc | $\longrightarrow$ | $OAc^-$ | + $H_2O$ |
|---|---|---|---|---|---|---|
| Before reaction | 25.0 × 0.10 *M* = 2.5 mmol | | 5.0 mL × 0.10 *M* = 5.0 mmol | | 0 | |
| After reaction | 2.5 − 2.5 = 0 | | 5.0 − 2.5 = 2.5 mmol | | 2.5 mmol | |

After the reaction goes to completion, the solution contains 2.5 mmol HOAc and 2.5 mmol $OAc^-$

4. After the reaction, the major species in solution are HOAc, $OAc^-$, $Na^+$, $H_2O$. The equilibrium that will control the pH is

$$HOAc \rightleftarrows H^+ + OAc^- \qquad K_a = \frac{[H^+][OAc^-]}{[HOAc]} = 1.8 \times 10^{-5}$$

| *Initial Concentrations (before any dissociation of HOAc)* | | *Equilibrium Concentrations* |
|---|---|---|
| $[HOAc]_0 = \dfrac{2.5\text{ mmol}}{(50.0 + 25.0)\text{mL}}$ | Let *x* mol/L HOAc dissociate ⟶ | $[HOAc] = \dfrac{2.5}{75} = x$ |
| $[OAc^-]_0 = \dfrac{2.5\text{ mmol}}{(50.0 + 25.0)\text{mL}}$ | | $[OAc^-] = \dfrac{2.5}{75} + x$ |
| $[H^+]_0 \approx 0$ | | $[H^+] = x$ |

$$1.8 \times 10^{-5} = K_a = \frac{[H^+][OAc^-]}{[HOAc]} = \frac{(x)\left(\frac{2.5}{75} + x\right)}{\left(\frac{2.5}{75} - x\right)} \approx \frac{x\left(\frac{2.5}{75}\right)}{\left(\frac{2.5}{75}\right)}$$

$$x = 1.8 \times 10^{-5} = [H^+]$$

$$pH = -\log[H^+] = -\log(1.8 \times 10^{-5}) = 4.74$$

## Test 3.3

**A.** 1. 25.0 mL $\times$ 0.30 $M$ $HNO_2$ = 7.5 mmol $HNO_2$. The titration reaction is

$$OH^- + HNO_2 \longrightarrow H_2O + NO_2^-$$

Thus 7.5 mmol of $OH^-$ will be required:

$$(x \text{ mL})(0.50\ M\ \text{NaOH}) = 7.5 \text{ mmol } OH^-$$

$$x = \frac{7.5}{0.50} = 15 \text{ mL}$$

2a. Before any reaction, the solution contains $HNO_2$, $Na^+$, $OH^-$, $H_2O$. The reaction will be

| | $HNO_2$ | + | $OH^-$ | $\longrightarrow$ | $NO_2^-$ | + $H_2O$ |
|---|---|---|---|---|---|---|
| Before reaction | 25.0 mL $\times$ 0.30 $M$ = 7.5 mmol | | 5.0 mL $\times$ 0.50 $M$ = 2.5 mmol | | 0 | |
| After reaction | 7.5 − 2.5 = 5.0 mmol | | 0 | | 2.5 mmol | |

After the reaction, the solution contains $HNO_2$, $NO_2^-$, $Na^-$, $H_2O$. The equilibrium that will control the $[H^+]$ is

$$HNO_2 \rightleftarrows H^+ + NO_2^- \qquad K_a = \frac{[H^+][NO_2^-]}{[HNO_2]} = 4.0 \times 10^{-4}$$

| *Initial Concentrations* | | *Equilibrium Concentrations* |
|---|---|---|
| $[HNO_2]_0 = \frac{5.0 \text{ mmol}}{(25.0 + 5.0)\text{mL}} = 1.7 \times 10^{-1}\ M$ | | $[HNO_2] = 1.7 \times 10^{-1} - x$ |
| $[NO_2^-]_0 = \frac{2.5 \text{ mmol}}{(25.0 + 5.0)\text{mL}} = 8.3 \times 10^{-2}\ M$ | Let $x$ mol/L $HNO_2$ dissociate $\longrightarrow$ | $[NO_2^-] = 8.3 \times 10^{-2} + x$ |
| $[H^+]_0 \approx 0$ | | $[H^+] = x$ |

$$4.0 \times 10^{-4} = K_a = \frac{[H^+][NO_2^-]}{[HNO_2]} = \frac{(x)(8.3 \times 10^{-2} + x)}{(1.7 \times 10^{-1} - x)} \approx \frac{(x)(8.3 \times 10^{-2})}{(1.7 \times 10^{-1})}$$

$$x \approx \frac{4.0 \times 10^{-4}(1.7 \times 10^{-1})}{8.3 \times 10^{-2}} = 8.2 \times 10^{-4}$$

Assumptions are valid. $[H^+] = 8.2 \times 10^{-4}$; pH = 3.09.

2b. At this point, 7.5 mL $\times$ 0.50 $M$ NaOH = 3.75 mmol of NaOH have been added. This will react with half of the 7.5 mmol of $HNO_2$ originally present. Thus, this point is halfway to the equivalence point, so

$$[HNO_2] \approx [NO_2^-]$$

$$K_a = \frac{[H^+][NO_2^-]}{[HNO_2]} = [H^+]$$

$$[H^+] = 4.0 \times 10^{-4};\ pH = 3.40$$

2c. This is the equivalence point: the 7.5 mmol $HNO_2$ have been changed to 7.5 mmol of $NO_2^-$. After the titration reaction, the solution contains $NO_2^-$, $Na^+$, $H_2O$. Note that $NO_2^-$ is a base ($HNO_2$ is a weak acid):

$$NO_2^- + H_2O \rightleftarrows HNO_2 + OH^-$$

$$K_b = \frac{[HNO_2][OH^-]}{[NO_2^-]} = \frac{K_w}{K_a} = \frac{1.00 \times 10^{-14}}{4.0 \times 10^{-4}} = 2.5 \times 10^{-11}$$

| *Initial Concentrations* | | *Equilibrium Concentrations* |
|---|---|---|
| $[NO_2^-]_0 = \frac{7.5\text{ mmol}}{(25.0 + 15.0)\text{mL}} = 1.9 \times 10^{-1}\ M$ | Let $x$ mol/L $NO_2^-$ react $\longrightarrow$ | $[NO_2^-] = 0.19 - x$ |
| $[HNO_2]_0 = 0$ | | $[HNO_2] = x$ |
| $[OH^-]_0 \approx 0$ | | $[OH^-] = x$ |

$$2.5 \times 10^{-11} = K_b = \frac{(x)(x)}{0.19 - x} \approx \frac{x^2}{0.19}$$

$$x^2 \approx 4.7 \times 10^{-12}$$

$$x \approx 2.2 \times 10^{-6}$$

Assumption is valid.

$$[OH^-] = x = 2.2 \times 10^{-6}$$

$$[H^+] = \frac{K_w}{[OH^-]} = \frac{1.0 \times 10^{-14}}{2.2 \times 10^{-6}} = 4.6 \times 10^{-9}$$

$$pH = 8.34$$

2d. This is 5.0 mL beyond the equivalence point. The solution contains $OH^-$, $Na^+$, $NO_2^-$, $H_2O$. The excess $OH^-$ will determine the pH.

$$[OH^-] = \frac{10.0\text{ mmol added} - 7.5\text{ mmol consumed}}{(25.0 + 20.0)\text{mL}}$$

$$= \frac{2.5}{45} = 5.6 \times 10^{-2}$$

$$[H^+] = \frac{K_w}{[OH^-]} = \frac{1.0 \times 10^{-14}}{5.6 \times 10^{-2}} = 1.8 \times 10^{-13}\ M$$

$$pH = 12.74$$

**B.** 1. 40.0 mL $\times$ 0.0500 $M$ NaOH = 2.00 mmol NaOH. The titration reaction is

$$OH^- + HA \longrightarrow A^- + H_2O$$

Since NaOH reacts with HA in a 1 : 1 ratio, the original solution must have contained 2.00 mmol of HA.

$$0.200\text{ g HA} = 2.00\text{ mmol HA} = 2.00 \times 10^{-3}\text{ mol HA}$$

$$\frac{0.20\text{ g HA}}{2.00 \times 10^{-3}\text{ mol HA}} = 1.00 \times 10^2\text{ g/mol}$$

The molecular weight of HA is 100 g/mol.

2. The titration reaction is

| | HA | + | $OH^-$ | $\longrightarrow$ | $H_2O$ + | $A^-$ |
|---|---|---|---|---|---|---|
| Before reaction | 2.00 mmol (see part 1) | | 20.0 mL × 0.0500 *M* = 1.00 mmol | | | 0 |
| After reaction | 2.00 − 1.00 = 1.00 mmol | | 1.00 − 1.00 = 0 | | | 1.00 mmol |

After the reaction, the solution contains HA, $A^-$, $Na^+$, $H_2O$. The pH will be determined by the reaction

$$HA \rightleftarrows H^+ + A^- \qquad K_a = \frac{[H^+][A^-]}{[HA]}$$

*Initial Concentrations* | *Equilibrium Concentrations*

$$[HA]_0 = \frac{1.00\text{ mmol}}{(100.0 + 20.0)\text{mL}} = 8.33 \times 10^{-3}\ M \qquad [HA] = 8.33 \times 10^{-3} - x$$

$$[A^-]_0 = \frac{1.00\text{ mmol}}{(100.0 + 20.0)\text{mL}} = 8.33 \times 10^{-3}\ M \qquad \xrightarrow[\text{dissociate}]{\text{Let } x \text{ mol/L HA}} \qquad [A^-] = 8.33 \times 10^{-3} + x$$

$$[H^+]_0 \approx 0 \qquad [H^+] = x$$

In this case, the pH = 6.0. Thus $[H^+] = 10^{-pH} = 1.0 \times 10^{-6} = x$.

$$K_a = \frac{[H^+][A^-]}{[HA]} = \frac{(1.0 \times 10^{-6})(8.33 \times 10^{-3} + 1.0 \times 10^{-6})}{(8.33 \times 10^{-3} - 1.0 \times 10^{-6})}$$

$$\approx \frac{(1.0 \times 10^{-6})(8.33 \times 10^{-3})}{(8.33 \times 10^{-3})}$$

$$= 1.0 \times 10^{-6}$$

Note that there is an easier way to think about this problem. Since the original solution contained 2.00 mmol of HA, and 20.0 mL of added 0.0500 *M* NaOH contains 1.00 mmol of $OH^-$, this is the halfway point in the titration where $[HA] \approx [A^-]$. (This is demonstrated above.) Thus,

$$[H^+] = K_a = 1.0 \times 10^{-6}$$

3. At the equivalence point (40.0 mL of added 0.0500 *M* NaOH), the solution contains $A^-$, $Na^+$, $H_2O$. $A^-$ is a base (HA is a weak acid) and will react with $H_2O$:

$$A^- + H_2O \rightleftarrows HA + OH^-$$

$$K_b = \frac{[HA][OH^-]}{[A^-]} = \frac{K_w}{K_a} = \frac{1.0 \times 10^{-14}}{1.0 \times 10^{-6}} = 1.0 \times 10^{-8}$$

*Initial Concentrations* | *Equilibrium Concentrations*

$$[A^-]_0 = \frac{2.0\text{ mmol}}{(100.0 + 40.0)\text{mL}} = 1.4 \times 10^{-2}\ M \qquad [A^-] = 1.4 \times 10^{-2} - x$$

$$[HA]_0 = 0 \qquad \xrightarrow[\text{react}]{\text{Let } x \text{ mol/L of } A^-} \qquad [HA] = x$$

$$[OH^-]_0 \approx 0 \qquad [OH^-] = x$$

$$1.0 \times 10^{-8} = K_b = \frac{[OH^-][HA]}{[A^-]} = \frac{(x)(x)}{(1.4 \times 10^{-2} - x)} \approx \frac{x^2}{1.4 \times 10^{-2}}$$

$$x^2 \approx 1.4 \times 10^{-10}$$

$$x \approx 1.2 \times 10^{-5}$$

Assumption is valid.

$$[OH^-] = 1.2 \times 10^{-5}$$

$$[H^+] = \frac{K_w}{[OH^-]} = \frac{1.0 \times 10^{-14}}{1.2 \times 10^{-5}} = 8.3 \times 10^{-10};\ pH = 9.08$$

**Test 3.4**

**A.** The major species in solution are $NH_3$ and $H_2O$. The dominant equilibrium is

$$NH_3 + H_2O \rightleftarrows NH_4^+ + OH^-$$

$$K_b = \frac{[NH_4^+][OH^-]}{[NH_3]} = 1.8 \times 10^{-5}$$

| *Initial Concentrations (before any $NH_3$ reacts with $H_2O$)* | | *Equilibrium Concentrations* |
|---|---|---|
| $[NH_3]_0 = 0.05\ M$ | | $[NH_3] = 0.050 - x$ |
| $[NH_4^+]_0 = 0$ | Let $x$ mol/L $NH_3$ react $\longrightarrow$ | $[NH_4^+] = x$ |
| $[OH^-]_0 \approx 0$ | | $[OH^-] = x$ |

$$1.8 \times 10^{-5} = K_b = \frac{[NH_4^+][OH^-]}{[NH_3]} = \frac{(x)(x)}{(0.050 - x)} \approx \frac{x^2}{0.050}$$

$$x^2 \approx (5.0 \times 10^{-2})(1.8 \times 10^{-5}) = 9.0 \times 10^{-7}$$

$$x \approx 9.5 \times 10^{-4}$$

$$[OH^-] = x = 9.5 \times 10^{-4}\ M$$

$$[H^+][OH^-] = 1.00 \times 10^{-14} = [H^+](9.5 \times 10^{-4})$$

$$[H^+] = \frac{1.00 \times 10^{-14}}{9.5 \times 10^{-4}} = 1.1 \times 10^{-11}\ M$$

$$pH = 10.98$$

**B.** 1. The stoichiometry problem.

| | $NH_3$ | + | $H^+$ | $\longrightarrow$ | $NH_4^+$ |
|---|---|---|---|---|---|
| Before reaction | 100.0 mL × 0.050 $M$ = 5.0 mmol | | 10.0 mL × 0.10$M$ = 1.0 mmol | | 0 |
| After reaction | 5.0 − 1.0 = 4.0 mmol | | 0 | | 1.0 mmol |

2. The equilibrium problem.

After the reaction, the solution contains $NH_3$, $NH_4^+$, $Cl^-$, $H_2O$. We must use an equilibrium that involves both $NH_3$ and $NH_4^+$. There are two choices:

$$NH_4^+ \rightleftarrows NH_3 + H^+ \qquad K_a$$

$$NH_3 + H_2O \rightleftarrows NH_4^+ + OH^- \qquad K_b$$

Either will give the correct answer. Let's use the dissociation equilibrium of $NH_4^+$ since it contains $H^+$, which will allow us to calculate the $[H^+]$ directly:

$$K_a = \frac{K_w}{K_b} = \frac{1.00 \times 10^{-14}}{1.8 \times 10^{-5}} = \frac{[NH_3][H^+]}{[NH_4^+]} = 5.6 \times 10^{-10}$$

*Initial Concentrations (before $NH_4^+$ dissociates)*

$$[NH_4^+]_0 = \frac{1.0 \text{ mmol}}{(100.0 + 10.0)\text{mL}}$$

$$[NH_3]_0 = \frac{4.0 \text{ mmol}}{(100.0 + 10.0)\text{mL}}$$

$$[H^+]_0 \approx 0$$

$\xrightarrow[\text{dissociate}]{\text{Let } x \text{ mol/L } NH_4^+}$

*Equilibrium Concentrations*

$$[NH_4^+] = \frac{1}{110} - x$$

$$[NH_3] = \frac{4}{110} + x$$

$$[H^+] = x$$

$$5.6 \times 10^{-10} = K_a = \frac{[H^+][NH_3]}{[NH_4^+]} = \frac{(x)\left(\frac{4}{110} + x\right)}{\left(\frac{1}{110} - x\right)} \approx \frac{x\left(\frac{4}{110}\right)}{\left(\frac{1}{110}\right)} = \frac{(x)(4)}{1}$$

$$x \approx \left(\frac{5.6 \times 10^{-10}}{4}\right) = 1.4 \times 10^{-10}\ M = [H^+]$$

$$pH = 9.85$$

**C.** 25.0 mL of 0.10 $M$ HCl has been added. 25.0 mL $\times$ 0.10 $M$ = 2.5 mmol $H^+$ have been added. Originally, there were 100.0 mL $\times$ 0.050 $M$ = 5.0 mmol $NH_3$. Thus, 25.0 mL of added 0.10 $M$ HCl represents the halfway point in this titration (half of the $NH_3$ has been converted to $NH_4^+$). At this point, $[NH_3] \approx [NH_4^+]$.

$$K_a = \frac{[H^+][NH_3]}{[NH_4^+]} = 5.6 \times 10^{-10}$$

$$[H^+] = 5.6 \times 10^{-10}\ M$$

$$pH = 9.25$$

**D.** 50.0 mL of added 0.10 $M$ HCl corresponds to the equivalence point of this titration: 5.0 mmol $H^+$ reacts with 5.0 mmol $NH_3$ to produce 5.0 mmol $NH_4^+$. After the titration reaction, the solution contains the major species $NH_4^+$, $Cl^-$, $H_2O$. The equilibrium that will dominate is

$$NH_4^+ \rightleftarrows NH_3 + H^+$$

$$K_a = 5.6 \times 10^{-10} = \frac{[NH_3][H^+]}{[NH_4^+]}$$

*Initial Concentrations (before any $NH_4^+$ dissociates)*

$$[NH_4^+]_0 = \frac{5.0 \text{ mmol}}{(100.0 + 50.0)\text{mL}}$$

$$[NH_3]_0 = 0$$

$$[H^+]_0 \approx 0$$

$\xrightarrow[\text{dissociate}]{\text{Let } x \text{ mol/L } NH_4^+}$

*Equilibrium Concentrations*

$$[NH_4^+] = \frac{5}{150} - x$$

$$[NH_3] = x$$

$$[H^+] = x$$

$$5.6 \times 10^{-10} = \frac{[NH_3][H^+]}{[NH_4^+]} = \frac{(x)(x)}{\left(\frac{5}{150} - x\right)} \approx \frac{x^2}{\frac{5}{150}} = \frac{x^2}{3.3 \times 10^{-2}}$$

$$x^2 \approx 1.87 \times 10^{-11}$$

$$x \approx 4.3 \times 10^{-6} = [H^+]$$

$$pH = -\log(4.3 \times 10^{-6}) = 5.36$$

**E.** 60.0 mL of added 0.10 $M$ HCl represents 10.0 mL beyond the equivalence point. Thus, the solution contains excess $H^+$.

$$60.0 \text{ mL} \times 0.10\ M = 6.0 \text{ mmol} \quad H^+ \text{ added}$$

$$100.0 \text{ mL} \times 0.050\ M = 5.0 \text{ mmol} \quad NH_3 \text{ originally present}$$

$$6.0 - 5.0 = 1.0 \text{ mmol } H^+ \text{ in excess}$$

$$[H^+] = \frac{1.0 \text{ mmol}}{160 \text{ mL}} = 6.3 \times 10^{-3}\ M$$

$$pH = 2.20$$

### Test 3.5

**A.** Yellow (The solution is very acidic, so the indicator is present as HIn.)

**B.** The normal assumption is that the color change will be visible when

$$\frac{[In^-]}{[HIn]} \approx \frac{1}{10}$$

$$K_a = 1.0 \times 10^{-8} = \frac{[H^+][In^-]}{[HIn]} = [H^+]\frac{1}{10}$$

$$[H^+] = (10)(1.0 \times 10^{-8}) = 1.0 \times 10^{-7}\ M$$

This is the $[H^+]$ when the color change should first be visible. pH = 7.0 at color change.

**C.** 300.0 mL of NaOH represents an excess. The solution will be blue. In a basic solution, the indicator will be predominantly in the $In^-$ form.

### Test 3.6

The normal assumption is that the color change will be visible when $[In^-]/[HIn] = 1/10$. For a color change at pH 5 ($[H^+] = 1.0 \times 10^{-5}$):

$$K_a = [H^+]\frac{[In^-]}{[HIn]} = (1.0 \times 10^{-5})\left(\frac{1}{10}\right) = 1.0 \times 10^{-6}$$

For a color change at pH 9 ($[H^+] = 1.0 \times 10^{-9}$):

$$K_a = [H^+]\frac{[In^-]}{[HIn]} = (1.0 \times 10^{-9})\left(\frac{1}{10}\right) = 1.0 \times 10^{-10}$$

Thus, any indicator with a $K_a$ value in the range from $10^{-6}$ to $10^{-10}$ will serve nicely to mark the endpoint of a typical titration of a strong acid with a strong base.

### Test 3.7

The titration reaction is

$$HCN + OH^- \longrightarrow CN^- + H_2O$$

The solution originally contains (100.0 mL)(0.100 $M$) = 10.0 mmol of HCN. This will be changed to 10.0 mmol of $CN^-$ at the equivalence point. (100.0 mL of 0.100 $M$ NaOH is required to reach the equivalence point.) At the equivalence point, the solution contains the major species $Na^+$, $CN^-$, $H_2O$. The equilibrium that will dominate is

$$CN^- + H_2O \rightleftarrows HCN + OH^-$$

$$K_b = \frac{[HCN][OH^-]}{[CN^-]} = \frac{K_w}{K_a} = \frac{1.0 \times 10^{-14}}{6.2 \times 10^{-10}} = 1.6 \times 10^{-5}$$

| *Initial Concentrations* | | *Equilibrium Concentrations* |
|---|---|---|
| $[CN^-]_0 = \dfrac{10.0\text{ mmol}}{(100.0 + 100.0)\text{mL}} = 5.00 \times 10^{-2}\ M$ | Let $x$ mol/L $CN^-$ react with $H_2O$ $\longrightarrow$ | $[CN^-] = 5.00 \times 10^{-2} - x$ |
| $[HCN]_0 = 0$ | | $[HCN] = x$ |
| $[OH^-]_0 \approx 0$ | | $[OH^-] = x$ |

$$1.6 \times 10^{-5} = K_b = \frac{[HCN][OH^-]}{[CN^-]} = \frac{(x)(x)}{(5.00 \times 10^{-2} - x)} \approx \frac{x^2}{5.00 \times 10^{-2}}$$

$$x^2 \approx 8.1 \times 10^{-7}$$

$$x \approx 9.0 \times 10^{-4}$$

Simplification is valid, so

$$[OH^-] = x = 9.0 \times 10^{-4}$$

$$[H^+][OH^-] = 1.00 \times 10^{-14} = [H^+](9.0 \times 10^{-4})$$

$$[H^+] = \frac{1.00 \times 10^{-14}}{9.0 \times 10^{-4}} = 1.11 \times 10^{-11}\ M$$

$$\text{pH} = 10.95$$

The pH at the equivalence point is 10.95. The indicator equilibrium is

$$HIn \rightleftarrows H^+ + In^-$$

$$K_a = [H^+]\left(\frac{[In^-]}{[HIn]}\right) = (1.1 \times 10^{-11})\left(\frac{1}{10}\right) = 1.1 \times 10^{-12}$$

The indicator should have a $K_a$ value near $10^{-12}$.

### Test 3.8

The indicator is of the form HIn:

$$HIn \rightleftarrows H^+ + In^-$$

Ammonia is a base, so the ammonia solution will be basic, which will cause the indicator to be predominantly in the $In^-$ form. The color change will be apparent when enough acid has been added, so that $[In^-]/[HIn] \approx 10/1$. The pH at the equivalence point is 5, so $[H^+] = 10^{-5}$.

$$K_a = \frac{[H^+][In^-]}{[HIn]} = [H^+]\left(\frac{[In^-]}{[HIn]}\right) = (10^{-5})\left(\frac{10}{1}\right) = 10^{-4}$$

$K_a$ for an appropriate indicator should be approximately $10^{-4}$.

---

## Chapter 4

### Test 4.1

**A.** $AgI(s) \rightleftarrows Ag^+ + I^-$

**B.** $K_{sp} = [Ag^+][I^-]$

**C.** $[Ag^+]_0 = 0$
$[I^-]_0 = 0$

**D.** $x$ = mol/L of $AgI(s)$ that dissolves to come to equilibrium.

**E.** $[Ag^+] = x$
$[I^-] = x$

**F.** In this case, $x$ = solubility = $1.2 \times 10^{-8}$ mol/L.

**G.** $K_{sp} = [Ag^+][I^-] = (x)(x) = (1.2 \times 10^{-8})(1.2 \times 10^{-8}) = 1.4 \times 10^{-16}$

### Test 4.2

The solubility equilibrium for $Bi_2S_3(s)$ is

$$Bi_2S_3(s) \rightleftarrows 2Bi^{3+} + 3S^{2-} \qquad K_{sp} = [Bi^{3+}]^2[S^{2-}]^3$$

| *Initial Concentrations* | | *Equilibrium Concentrations* |
|---|---|---|
| $[Bi^{3+}]_0 = 0$ | Let $x$ mol/L $Bi_2S_3$ dissolve $\longrightarrow$ | $[Bi^{3+}] = 2x$ |
| $[S^{2-}]_0 = 0$ | | $[S^{2-}] = 3x$ |

Note: $x\ Bi_2S_3 \longrightarrow 2x\ Bi^{3+} + 3x\ S^{2-}$.

$$\text{Solubility} = x = 1.0 \times 10^{-15}\ \text{mol/L}$$
$$[Bi^{3+}] = 2x = 2(1.0 \times 10^{-15}) = 2.0 \times 10^{-15}\ \text{mol/L}$$
$$[S^{2-}] = 3x = 3(1.0 \times 10^{-15}) = 3.0 \times 10^{-15}\ \text{mol/L}$$
$$K_{sp} = [Bi^{3+}]^2[S^{2-}]^3 = (2.0 \times 10^{-15})^2(3.0 \times 10^{-15})^3$$
$$= (4.0 \times 10^{-30})(27 \times 10^{-45}) = 1.1 \times 10^{-73}$$

### Test 4.3

The equilibrium that occurs when $Ag_2CrO_4(s)$ dissolves is

$$Ag_2CrO_4(s) \rightleftarrows 2Ag^+(aq) + CrO_4^{2-}(aq)$$
$$K_{sp} = [Ag^+]^2[CrO_4^{2-}] = 9.0 \times 10^{-12}$$

| *Initial Concentrations* | | *Equilibrium Concentrations* |
|---|---|---|
| $[Ag^+]_0 = 0$ | Let $x$ mol/L $Ag_2CrO_4(s)$ dissolve $\longrightarrow$ | $[Ag^+] = 2x$ |
| $[CrO_4^{2-}]_0 = 0$ | | $[CrO_4^{2-}] = x$ |

Note: $x\ Ag_2CrO_4(s) \longrightarrow 2x\ Ag^+ + x\ CrO_4^{2-}$

$$9.0 \times 10^{-12} = K_{sp} = [Ag^+]^2[CrO_4^{2-}] = (2x)^2(x) = 4x^3$$
$$x^2 = \frac{9.0 \times 10^{-12}}{4} = 2.25 \times 10^{-12}$$
$$x = \sqrt[3]{2.25 \times 10^{-12}} = 1.3 \times 10^{-4}\ \text{mol/L} = \text{solubility}$$

### Test 4.4

*Solubility of CuS(s)*

$CuS(s) \rightleftarrows Cu^{2+} + S^{2-}$
$K_{sp} = [Cu^{2+}][S^{2-}] = 8.5 \times 10^{-45}$

Let $x$ = solubility. At equilibrium: $[Cu^{2+}] = x$ and $[S^{2-}] = x$, so

$$8.5 \times 10^{-45} = [Cu^{2+}][S^{2-}] = (x)(x)$$
$$x^2 = 8.5 \times 10^{-45}$$
$$x = 9.2 \times 10^{-23}\ \text{mol/L} = \text{solubility}$$

*Solubility of $Ag_2S(s)$*

$Ag_2S(s) \rightleftarrows 2Ag^+ + S^{2-}$
$K_{sp} = [Ag^+]^2[S^{2-}] = 1.6 \times 10^{-49}$

Let $x$ = solubility.

$x\ Ag_2S(s) \longrightarrow 2x\ Ag^+ + x\ S^{2-}$

At equilibrium: $[Ag^+] = 2x$ and $[S^{2-}] = x$, so

$$1.6 \times 10^{-49} = [Ag^+]^2[S^{2-}] = (2x)^2(x)$$
$$4x^3 = 1.6 \times 10^{-49}$$
$$x^3 = \frac{1.6 \times 10^{-49}}{4} = 4.0 \times 10^{-50}$$
$$x = 3.4 \times 10^{-17} \text{ mol/L} = \text{solubility}$$

*Solubility of $Bi_2S_3(s)$*

$Bi_2S_3(s) \rightleftarrows 2Bi^{3+} + 3S^{2-}$
$K_{sp} = [Bi^{3+}]^2[S^{2-}]^3 = 1.1 \times 10^{-73}$
Let $x$ = solubility.
$x\ Bi_2S_3(s) \longrightarrow 2x\ Bi^{3+} + 3x\ S^{2-}$
At equilibrium: $[Bi^{3+}] = 2x$ and $[S^{2-}] = 3x$, so

$$1.1 \times 10^{-73} = [Bi^{3+}]^2[S^{2-}]^3 = (2x)^2(3x)^3 = 108\ x^5$$
$$x^5 = \frac{1.1 \times 10^{-73}}{1.1 \times 10^{-2}} = 1.0 \times 10^{-75}$$
$$x = 1.0 \times 10^{-15} \text{ mol/L} = \text{solubility}$$

The solubilities of the three salts are

| *Salt* | $K_{sp}$ | *Solubility (mol/L)* |
|---|---|---|
| CuS | $8.3 \times 10^{-45}$ | $9.2 \times 10^{-23}$ |
| $Ag_2S$ | $1.6 \times 10^{-49}$ | $3.4 \times 10^{-17}$ |
| $Bi_2S_3$ | $1.1 \times 10^{-73}$ | $1.0 \times 10^{-15}$ |

The order of solubilities is

$$\underset{\text{most soluble}}{Bi_2S_3(s)} > Ag_2S(s) > \underset{\text{least soluble}}{CuS(s)}$$

This is the reverse of the order of the $K_{sp}$ values. This is an important result. The $K_{sp}$ values for salts that produce different numbers of ions (and thus which have different powers in the equilibrium constant expression) cannot be used directly to predict relative solubilities.

**Test 4.5**

*Solubility of $Ag_3PO_4(s)$ in Water* The equilibrium is

$$Ag_3PO_4(s) \rightleftarrows 3Ag^+ + PO_4^{3-}$$

And the equilibrium expression is

$$K_{sp} = [Ag^+]^3[PO_4^{3-}] = 1.8 \times 10^{-18}$$

| *Initial Concentrations* | | *Equilibrium Concentrations* |
|---|---|---|
| $[Ag^+]_0 = 0$ | Let $x$ mol/L $Ag_3PO_4(s)$ dissolve $\longrightarrow$ | $[Ag^+] = 3x$ |
| $[PO_4^{3-}]_0 = 0$ | | $[PO_4^{3-}] = x$ |

$$1.8 \times 10^{-18} = K_{sp} = [Ag^+]^3[PO_4^{3-}] = (3x)^3(x)$$
$$27x^4 = 1.8 \times 10^{-18}$$
$$x^4 = \frac{1.8 \times 10^{-18}}{27} = 6.7 \times 10^{-20}$$
$$x = 1.6 \times 10^{-5} \text{ mol/L} = \text{solubility}$$

*Solubility of $Ag_3PO_4(s)$ in 1.0 M $Na_3PO_4$* The equilibrium is

$$Ag_3PO_4(s) \rightleftarrows 3Ag^+ + PO_4^{3-}$$

and the equilibrium expression is

$$K_{sp} = [Ag^+]^3[PO_4^{3-}] = 1.8 \times 10^{-18}$$

| *Initial Concentrations (before any $Ag_3PO_4(s)$ dissolves)* | | *Equilibrium Concentrations* |
|---|---|---|
| $[Ag^+]_0 = 0$ | Let $x$ mol/L $Ag_3PO_4(s)$ dissolve $\longrightarrow$ | $[Ag^+] = 3x$ |
| $[PO_4^{3-}]_0 = 1.0\ M$ (from 1.0 $M$ $Na_3PO_4$) | | $[PO_4^{3-}] = 1.0 + x$ |

$$1.8 \times 10^{-18} = K_{sp} = [Ag^+]^3[PO_4^{3-}] = (3x)^3(1.0 + x)$$

Assume that $1.0 \gg x$ and thus that $1.0 + x \approx 1.0$ gives

$$1.8 \times 10^{-18} \approx (3x)^3(1.0)$$
$$27x^3 \approx 1.8 \times 10^{-18}$$
$$x^3 \approx 6.7 \times 10^{-20}$$
$$x \approx 4.1 \times 10^{-7} \text{ mol/L}$$

The assumption is valid. Solubility $= 4.1 \times 10^{-7}$ mol/L. Note that $Ag_3PO_4$ is significantly less soluble in 1.0 $M$ $Na_3PO_4$ than in pure water because of the common ion effect.

### Test 4.6

The mixed solution contains the ions $Cu^+$, $NO_3^-$, $Na^+$, $Cl^-$. We must first calculate the concentrations of $Cu^+$ and $Cl^-$ in the mixed solution, which has a volume of 250.0 mL:

$$[Cu^+]_0 = \frac{\text{mmol } Cu^+}{\text{mL of solution}} = \frac{(50.0 \text{ mL})(1.0 \times 10^{-2}\ M)}{(250.0 \text{ mL})} = 2.0 \times 10^{-3}\ M$$

$$[Cl^-]_0 = \frac{\text{mmol } Cl^-}{\text{mL of solution}} = \frac{(200.0 \text{ mL})(1.0 \times 10^{-4}\ M)}{(250.0 \text{ mL})} = 8.0 \times 10^{-5}\ M$$

The solid of interest here is $CuCl(s)$, which has the following ion product:

$$Q = [Cu^+]_0[Cl^-]_0$$
$$Q = (2.0 \times 10^{-3})(8.0 \times 10^{-5}) = 1.6 \times 10^{-7}$$
$$K_{sp} = 1.8 \times 10^{-7}$$
$$Q < K_{sp}$$

No precipitation of $CuCl(s)$.

### Test 4.7

First consider formation of $AgCl(s)$. The ion product expression is

$$Q_{AgCl} = [Ag^+]_0[Cl^-]_0$$

where $[Ag^+]_0 = 1.0 \times 10^{-4}\ M$. Let $Q = K_{sp} = 1.6 \times 10^{-10} = (1.0 \times 10^{-4})[Cl^-]_0$; then

$$[Cl^-]_0 = \frac{1.6 \times 10^{-10}}{1.0 \times 10^{-4}} = 1.6 \times 10^{-6}\ M$$

$AgCl(s)$ will form when $[Cl^-] > 1.6 \times 10^{-6}\ M$.

Consider $PbCl_2$. The ion product expression is

$$Q = [Pb^{2+}]_0[Cl^-]_0^2$$

where $[Pb^{2+}]_0 = 1.0 \times 10^{-2}\ M$. Let $Q = K_{sp} = 1.0 \times 10^{-4} = (1.0 \times 10^{-2})[Cl^-]_0^2$; then

$$[Cl^-]_0^2 = \frac{1.0 \times 10^{-4}}{1.0 \times 10^{-2}} = 1.0 \times 10^{-2}$$

$$[Cl^-]_0 = 1.0 \times 10^{-1}\ M$$

$PbCl_2(s)$ will form when $[Cl^-] > 1.0 \times 10^{-1}\ M$. As $Cl^-$ is added, $AgCl(s)$ will form first, when $[Cl^-] > 1.6 \times 10^{-6}\ M$.

**Test 4.8**

$pH = 9.00$

$[H^+] = 10^{-pH} = 1.00 \times 10^{-9}$

$$[S^{2-}] = \frac{1.3 \times 10^{-21}}{[H^+]^2} = \frac{1.3 \times 10^{-21}}{(1.00 \times 10^{-9})^2} = \frac{1.3 \times 10^{-21}}{1.0 \times 10^{-18}} = 1.3 \times 10^{-3}\ M$$

**Test 4.9**

The pH is 1.00; thus $[H^+] = 1.00 \times 10^{-1}\ M$. Since the solution is saturated with $H_2S$,

$$[S^{2-}] = \frac{1.3 \times 10^{-21}}{[H^+]^2} = \frac{1.3 \times 10^{-21}}{(1.00 \times 10^{-1})^2} = 1.3 \times 10^{-19}$$

Calculate $Q$ for each salt:

FeS($s$): $Q = [Fe^{2+}]_0[S^{2-}]_0 = (1.0 \times 10^{-2})(1.3 \times 10^{-19}) = 1.3 \times 10^{-21}$
$Q < K_{sp}$, so no FeS($s$) forms.

NiS($s$): $Q = [Ni^{2+}]_0[S^{2-}]_0 = (1.0 \times 10^{-2})(1.3 \times 10^{-19}) = 1.3 \times 10^{-21}$
$Q > K_{sp}$, so NiS($s$) forms.

---

## Chapter 5

### Test 5.1

**A.** $HNO_2 \rightleftarrows H^+ + NO_2^-$ $\quad K_a = \frac{[H^+][NO_2^-]}{[HNO_2]} = 4.0 \times 10^{-4}$

**B.** $[HNO_2]_0 = 2.00 \times 10^{-3}\ M$
$[NO_2^-]_0 = 0$
$[H^+]_0 \approx 0$ (ignore water as a contributor to $[H^+]$)

**C.** $x$ = mol/L of $HNO_2$ that dissociate

**D.** $[HNO_2] = 2.00 \times 10^{-3} - x$
$[NO_2^-] = x$
$[H^+] = x$

**E.** $4.0 \times 10^{-4} = K_a = \frac{[H^+][NO_2^-]}{[HNO_2]} = \frac{(x)(x)}{2.00 \times 10^{-3} - x}$

**F.** Assume $2.00 \times 10^{-3} - x \approx 2.00 \times 10^{-3}$:

$$\frac{x^2}{2.00 \times 10^{-3}} \approx 4.0 \times 10^{-4}$$

$$x^2 \approx 8.0 \times 10^{-7}$$

$$x \approx 8.9 \times 10^{-4}$$

**G.** Compare $x$ to $2.00 \times 10^{-3}$:

$$\frac{8.9 \times 10^{-4}}{2.00 \times 10^{-3}} \times 100 = 45\%$$

This clearly violates the 5% rule. That is,

$$2.00 \times 10^{-3} - x \neq 2.00 \times 10^{-3}$$

Thus the quadratic formula must be used.

**H.** $$\frac{x^2}{2.00 \times 10^{-3} - x} = 4.0 \times 10^{-4}$$

$$x^2 = (2.00 \times 10^{-3} - x)(4.0 \times 10^{-4}) = 8.0 \times 10^{-7} - (4.0 \times 10^{-4})x$$

$$x^2 + (4.0 \times 10^{-4})x - 8.0 \times 10^{-7} = 0$$

$a = 1$, $b = 4.0 \times 10^{-4}$, $c = -8.0 \times 10^{-7}$

$$x = \frac{-b \pm \sqrt{b^2 - 4ac}}{2a} =$$

$$\frac{-4.0 \times 10^{-4} \pm \sqrt{(4.0 \times 10^{-4})^2 - 4(1)(-8.0 \times 10^{-7})}}{2(1)}$$

$x = 7.0 \times 10^{-4}$ (ignore the negative root)

$[H^+] = x = 7.0 \times 10^{-4}$

pH = 3.15

### Test 5.2

**A.** Since the $HNO_3$ added produces $2 \times 10^{-8}$ $M$ $H^+$, which is close to the amount produced by water, both sources of $H^+$ must be taken into account. Use the charge balance equation:

$$[H^+] = [NO_3^-] + [OH^-]$$

$$[NO_3^-] = 2.0 \times 10^{-8}\ M$$

$$[OH^-] = \frac{K_w}{[H^+]}$$

$$[H^+] = [NO_3^-] + \frac{K_w}{[H^+]} = 2.0 \times 10^{-8} + \frac{1.0 \times 10^{-14}}{[H^+]}$$

$$[H^+] - \frac{1.0 \times 10^{-14}}{[H^+]} = 2.0 \times 10^{-8}$$

Using the quadratic formula,

$$[H^+] = 1.1 \times 10^{-7}\ M$$

$$pH = -\log(1.1 \times 10^{-7}) = 6.96$$

**B.** 1. The ions present are $Na^+$, $OH^-$, $H^+$.

2. The charge balance equation is

$[Na^+] + [H^+] = [OH^-]$

3. $[Na^+] = 1.0 \times 10^{-7}$ $M$ (from $1.0 \times 10^{-7}$ $M$ NaOH)

4. $[H^+][OH^-] = K_w; [H^+] = \dfrac{K_w}{[OH^-]}$

5. For this solution

$$[Na^+] + [H^+] = [OH^-] = [Na^+] + \frac{K_w}{[OH^-]}$$

$$[OH^-] - \frac{K_w}{[OH^-]} = [Na^+] = 1.0 \times 10^{-7}\ M$$

$$[OH^-] - \frac{1.0 \times 10^{-14}}{[OH^-]} = 1.0 \times 10^{-7}\ M$$

By trial and error: $[OH^-] = 1.6 \times 10^{-7}$

$[H^+][OH^-] = K_w = [H^+](1.6 \times 10^{-7}) = 1.0 \times 10^{-14}$

$$[H^+] = \frac{1.0 \times 10^{-14}}{1.6 \times 10^{-7}} = 6.3 \times 10^{-8}$$

$pH = 7.20$

The quadratic formula may also be used.

C. The HCl solution contains the ions $H^+$, $Cl^-$, and $OH^-$. The charge balance equation is

$$[H^+] = [Cl^-] + [OH^-]$$

In this case, $[Cl^-] = 0.10\ M$ (from the HCl). Since the $[OH^-]$ will be smaller than $10^{-7}\ M$ (the acid shifts the water equilibrium to the left),

$$[Cl^-] \gg [OH^-]$$

$$[H^+] = [Cl^-] + [OH^-] \approx [Cl^-] = 0.10\ M$$

$$pH = 1$$

This answer was obtained by ignoring the contribution of water to the $[H^+]$, since it is negligible in this case.

**Test 5.3**

$$6.2 \times 10^{-10} = K_a = \frac{[H^+]^2 - K_w}{[HA]_0 - \dfrac{[H^+]^2 - K_w}{[H^+]}}$$

$$= \frac{(2.5 \times 10^{-5})^2 - 1.00 \times 10^{-14}}{(1.0) - \dfrac{(2.5 \times 10^{-5})^2 - 1.00 \times 10^{-14}}{2.5 \times 10^{-5}}}$$

$$= \frac{6.25 \times 10^{-10} - 1.00 \times 10^{-14}}{(1.0) - \dfrac{6.25 \times 10^{-10} - 1.00 \times 10^{-14}}{2.5 \times 10^{-5}}} = \frac{6.25 \times 10^{-10}}{1.0 - \dfrac{6.25 \times 10^{-10}}{2.5 \times 10^{-5}}}$$

$$= \frac{6.25 \times 10^{-10}}{1.0}$$

The two sides of the equation are equal within round-off errors. In this problem, the full equation and the "normal equation" give the same answer. It is not necessary to consider water as a source of $H^+$ here. The 1.0 $M$ HCN produces much more $H^+$ than does $H_2O$.

### Test 5.4

**A.** The dissociation equilibrium for phenol is

$$C_6H_5OH \rightleftarrows C_6H_5O^- + H^+$$

$$K_a = \frac{[H^+][C_6H_5O^-]}{[C_6H_5OH]} = 1.3 \times 10^{-10}$$

Doing the problem using the normal steps leads to the expression

$$1.3 \times 10^{-10} = \frac{(x)(x)}{2.0 \times 10^{-4} - x} \approx \frac{x^2}{2.0 \times 10^{-4}}$$

$$x^2 \approx 2.6 \times 10^{-14}$$

$$x \approx 1.6 \times 10^{-7}\ M = [H^+] \quad \text{from phenol without considering water}$$

**B.** $1.6 \times 10^{-7}$ $M$ is less than $10^{-6}$ $M$. Thus water must be considered in calculating the correct value of the $[H^+]$.

**C.** The full equation must be used.

**D.** The $[H^+]$ from phenol considered alone is $1.6 \times 10^{-7}$ $M$. Water will contribute something to this. A reasonable guess might be $[H^+] = 2.0 \times 10^{-7}$ $M$.

**E.** $$K_a = 1.3 \times 10^{-10} = \frac{[H^+]^2 - K_w}{[HA]_0 - \dfrac{[H^+]^2 - K_w}{[H^+]}}$$

$$= \frac{[H^+]^2 - 1.0 \times 10^{-14}}{2.0 \times 10^{-4} - \dfrac{[H^+]^2 - 1.0 \times 10^{-14}}{[H^+]}}$$

Use successive approximations. Substitute guessed value of $2.0 \times 10^{-7}$ $M$ for $[H^+]$ into the denominator:

$$1.3 \times 10^{-10} = \frac{[H^+]^2 - 1.0 \times 10^{-14}}{2.0 \times 10^{-4} - \dfrac{(2.0 \times 10^{-7})^2 - 1.0 \times 10^{-14}}{2.0 \times 10^{-7}}}$$

$$= \frac{[H^+]^2 - 1.0 \times 10^{-14}}{2.0 \times 10^{-4} - 1.5 \times 10^{-7}}$$

Solving for $[H^+]$ gives

$$[H^+] = 1.9 \times 10^{-7}\ M$$

Substituting this value back into the equation shows that it is the correct value.

### Test 5.5

**A.** Since $K_{a_1} \gg K_{a_2}$, assume that only $H_2CO_3$ is a significant source of $H^+$ in this solution. This assumption allows the treatment of this acid to be the same as for a normal monoprotic acid:

$$H_2CO_3 \rightleftarrows H^+ + HCO_3^-$$

$$K_{a_1} = \frac{[H^+][HCO_3^-]}{[H_2CO_3]} = 4.3 \times 10^{-7}$$

*Initial Concentrations* | *Equilibrium Concentrations*

$[H_2CO_3]_0 = 0.10\ M$ $\xrightarrow[\text{dissociate}]{\text{Let } x \text{ mol/L } H_2CO_3}$ $[H_2CO_3] = 0.10 - x$

$[HCO_3^-]_0 = 0$ $\qquad$ $[HCO_3^-] = x$

$[H^+]_0 \approx 0$ $\qquad$ $[H^+] = x$

$$4.3 \times 10^{-7} = \frac{[H^+][HCO_3^-]}{[H_2CO_3]} = \frac{(x)(x)}{0.10 - x} \approx \frac{x^2}{0.10}$$

$$x^2 \approx (0.10)(4.3 \times 10^{-7}) = 4.3 \times 10^{-8}$$

$$x \approx 2.1 \times 10^{-4} = [H^+]$$

$$pH = 3.68$$

**B.** Use $K_{a_2}$:

$$5.6 \times 10^{-11} = K_{a_2} = \frac{[CO_3^{2-}][H^+]}{[HCO_3^-]}$$

From part A,

$$[H^+] = 2.1 \times 10^{-4}\ M$$

$$[HCO_3^-] = 2.1 \times 10^{-4}\ M$$

$$5.6 \times 10^{-11} = \frac{[CO_3^{2-}][H^+]}{[HCO_3^-]} = \frac{[CO_3^{2-}](2.1 \times 10^{-4})}{2.1 \times 10^{-4}} = [CO_3^{2-}]$$

$$[CO_3^{2-}] = 5.6 \times 10^{-11}\ M$$

**Test 5.6**

$H_2SO_4 \rightleftarrows H^+ + HSO_4^-$ $\qquad$ $K_a$ large

$HSO_4^- \rightleftarrows H^+ + SO_4^{2-}$ $\qquad$ $K_a = 1.2 \times 10^{-2}$

The major species in 5.0 $M$ $H_2SO_4$ are

$H^+$, $HSO_4^-$, $H_2O$

from complete dissociation of $H_2SO_4$

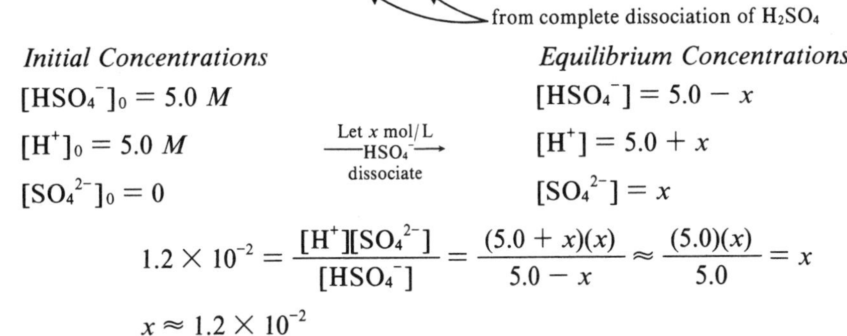

*Initial Concentrations* | *Equilibrium Concentrations*

$[HSO_4^-]_0 = 5.0\ M$ $\xrightarrow[\text{dissociate}]{\text{Let } x \text{ mol/L } HSO_4^-}$ $[HSO_4^-] = 5.0 - x$

$[H^+]_0 = 5.0\ M$ $\qquad$ $[H^+] = 5.0 + x$

$[SO_4^{2-}]_0 = 0$ $\qquad$ $[SO_4^{2-}] = x$

$$1.2 \times 10^{-2} = \frac{[H^+][SO_4^{2-}]}{[HSO_4^-]} = \frac{(5.0 + x)(x)}{5.0 - x} \approx \frac{(5.0)(x)}{5.0} = x$$

$$x \approx 1.2 \times 10^{-2}$$

Assumption is valid.

$[H^+] = 5.0$

$pH = -\log(5.0) = -0.70$

$[SO_4^{2-}] = x = 1.2 \times 10^{-2}\ M$

**Test 5.7**

**A.** The major species in solution are $Na^+$, $H_2AsO_4^-$, $H_2O$. The pH will be controlled by $H_2AsO_4^-$ acting both as an acid and as a base. Thus

$$pH = \frac{pK_1 + pK_2}{2} = \frac{2.30 + 7.08}{2} = \frac{9.38}{2} = 4.69$$

**B.** The major species in solution are $Na^+$, $HAsO_4^{2-}$, $H_2O$. The pH will be controlled by $HAsO_4^{2-}$ acting both as an acid and as a base.

$$pH = \frac{pK_2 + pK_3}{2} = \frac{7.08 + 9.22}{2} = \frac{16.30}{2} = 8.15$$

**C.** The major species in solution are $Na^+$, $H_2AsO_4^-$, $HAsO_4^{2-}$, $H_2O$. The strongest acid is $H_2AsO_4^-$. The strongest base is $HAsO_4^{2-}$. The dominant equilibrium is

$$H_2AsO_4^- \rightleftarrows H^+ + HAsO_4^{2-}$$

$$K_{a_2} = \frac{[H^+][HAsO_4^{2-}]}{[H_2AsO_4^-]} = 8.3 \times 10^{-8}$$

| *Initial Concentrations* | | *Equilibrium Concentrations* |
|---|---|---|
| $[H_2AsO_4^-]_0 = 0.100\ M$ | | $[H_2AsO_4^-] = 0.100 - x$ |
| $[HAsO_4^{2-}]_0 = 0.100\ M$ | Let $x$ mol/L $H_2AsO_4^-$ dissociate → | $[HAsO_4^{2-}] = 0.100 + x$ |
| $[H^+]_0 \approx 0$ | | $[H^+] = x$ |

$$8.3 \times 10^{-8} = K_{a_2} = \frac{[H^+][HAsO_4^{2-}]}{[H_2AsO_4^-]} = \frac{(x)(0.100 + x)}{0.100 - x} \approx \frac{(x)(0.100)}{0.100}$$

$$x = 8.3 \times 10^{-8} = [H^+]$$

$$pH = -\log(8.3 \times 10^{-8}) = 7.08$$

**Test 5.8**

The major species in solution are $H_2C_8H_4O_4$ and $H_2O$. $H_2C_8H_4O_4$ is a diprotic acid, but since $K_{a_1} \gg K_{a_2}$, assume that the stronger acid dominates:

$$H_2C_8H_4O_4 \rightleftarrows H^+ + HC_8H_4O_4^-$$

$$K_{a_1} = \frac{[H^+][HC_8H_4O_4^-]}{[H_2C_8H_4O_4]} = 1.3 \times 10^{-3}$$

| *Initial Concentrations* | | *Equilibrium Concentrations* |
|---|---|---|
| $[H_2C_8H_4O_4]_0 = 1.00\ M$ | | $[H_2C_8H_4O_4] = 1.00 - x$ |
| $[HC_8H_4O_4^-]_0 = 0$ | Let $x$ mol/L $H_2C_8H_4O_4$ dissociate → | $[HC_8H_4O_4^-] = x$ |
| $[H^+]_0 \approx 0$ | | $[H^+] = x$ |

$$1.3 \times 10^{-3} = \frac{[H^+][HC_8H_4O_4^-]}{[H_2C_8H_4O_4]} = \frac{(x)(x)}{1.00 - x} \approx \frac{x^2}{1.00}$$

$$x^2 \approx (1.00)(1.3 \times 10^{-3})$$

$$x \approx 3.6 \times 10^{-2}$$

$$\frac{3.6 \times 10^{-2}}{1.00} \times 100 = 3.6\% \quad \text{so the assumption is valid.}$$

$$x = [H^+] = 3.6 \times 10^{-2}$$

$$pH = -\log(3.6 \times 10^{-2}) = 1.44$$

**Test 5.9**

$$H_2C_8H_4O_4 \rightleftarrows H^+ + HC_8H_4O_4^-$$

$$K_{a_1} = \frac{[H^+][HC_8H_4O_4^-]}{[H_2C_8H_4O_4]} = 1.3 \times 10^{-3}$$

*Initial Concentrations* — *Equilibrium Concentrations*

$$[H_2C_8H_4O_4]_0 = \frac{75.0 \text{ mmol}}{125.0 \text{ mL}} = 0.600\ M \qquad [H_2C_8H_4O_4] = 0.600 - x$$

$$\xrightarrow[\text{dissociate}]{\text{Let } x \text{ mol/L } H_2C_8H_4O_4}$$

$$[HC_8H_4O_4^-]_0 = \frac{25.0 \text{ mmol}}{125.0 \text{ mL}} = 0.200\ M \qquad [HC_8H_4O_4^-] = 0.200 + x$$

$$[H^+]_0 \approx 0 \qquad [H^+] = x$$

$$1.3 \times 10^{-3} = K_{a_1} = \frac{[H^+][HC_8H_4O_4^-]}{[H_2C_8H_4O_4]} = \frac{(x)(0.200 + x)}{0.600 - x} \approx \frac{(x)(0.200)}{0.600}$$

$$x \approx \frac{(0.600)}{(0.200)}(1.3 \times 10^{-3}) = 3.9 \times 10^{-3}$$

Assumption is valid.

$[H^+] = 3.9 \times 10^{-3}$

$pH = -\log(3.9 \times 10^{-3}) = 2.41$

**Test 5.10**

The reaction is

| | $OH^-$ | + $H_2C_8H_4O_4$ | $\longrightarrow$ $HC_8H_4O_4^-$ | + $H_2O$ |
|---|---|---|---|---|
| Before reaction | 50.0 mmol | 100.0 mmol | 0 | |
| After reaction | 0 | 50.0 mmol | 50.0 mmol | |

When the reaction has gone to completion, the solution contains (major species)

$$H_2C_8H_4O_4,\ HC_8H_4O_4^-,\ Na^+,\ H_2O$$

The equilibrium that controls the pH will involve both $H_2C_8H_4O_4$ and $HC_8H_4O_4^-$:

*Initial Concentrations* — *Equilibrium Concentrations*

$$[H_2C_8H_4O_4]_0 = \frac{50.0}{150.} = 0.333\ M \qquad [H_2C_8H_4O_4] = 0.333 - x$$

$$\xrightarrow[\text{dissociate}]{\text{Let } x \text{ mol/L } H_2C_8H_4O_4}$$

$$[HC_8H_4O_4^-]_0 = \frac{50.0}{150.} = 0.333\ M \qquad [HC_8H_4O_4^-] = 0.333 + x$$

$$[H^+]_0 \approx 0 \qquad [H^+] = x$$

$$1.3 \times 10^{-3} = K_{a_1} = \frac{[H^+][HC_8H_4O_4^-]}{[H_2C_8H_4O_4]} = \frac{(x)(0.333 + x)}{0.333 - x} \approx \frac{(x)(0.333)}{0.333}$$

$$x = 1.3 \times 10^{-3}\ M = [H^+] = K_{a_1}$$

$$pH = pK_{a_1} = 2.89$$

Note that this is halfway to the first stoichiometric point, where $pH = pK_{a_1}$.

**Test 5.11**

At this point, $(150.0 \text{ mL})(1.00\ M) = 150.0$ mmol of $OH^-$ have been added. This will react with the 100.0 mmol of $H_2C_8H_4O_4$ to produce 50.0 mmol of $C_8H_4O_4^{2-}$ and 50.0

mmol of $HC_8H_4O_4^-$. When the reaction with $OH^-$ is complete, the solution contains

$$HC_8H_4O_4^-, C_8H_4O_4^{2-}, Na^+, H_2O$$

As in part E, $K_{a_2}$ must be used to solve for the $[H^+]$.

Take a shortcut by noting that this point in the titration corresponds to the halfway point between the first and second equivalence points (the amounts of $HC_8H_4O_4^-$ and $C_8H_4O_4^{2-}$ are equal). At this point

$$pH = pK_{a_2} = -\log(3.9 \times 10^{-6}) = 5.41$$

**Test 5.12**

The solution contains $Na^+$, $C_8H_4O_4^{2-}$, $H_2O$. The dominant reaction will be

$$C_8H_4O_4^{2-} + H_2O \rightleftarrows HC_8H_4O_4^- + OH^-$$

$$K_b = \frac{K_w}{K_{a_2}} = \frac{1.0 \times 10^{-14}}{3.9 \times 10^{-6}} = 2.6 \times 10^{-9}$$

*Initial Concentrations*

$$[C_8H_4O_4^{2-}]_0 = \frac{100.0 \text{ mmol}}{300.0 \text{ mL}} = 0.333\ M$$

$$[HC_8H_4O_4^-]_0 = 0$$

$$[OH^-]_0 \approx 0$$

Let $x$ mol/L $C_8H_4O_4^{2-}$ react with $H_2O$ $\longrightarrow$

*Equilibrium Concentrations*

$$[C_8H_4O_4^{2-}] = 0.333 - x$$

$$[HC_8H_4O_4^-] = x$$

$$[OH^-] = x$$

$$2.6 \times 10^{-9} = K_b = \frac{[HC_8H_4O_4^-][OH^-]}{[C_8H_4O_4^{2-}]} = \frac{(x)(x)}{0.333 - x} \approx \frac{x^2}{0.333}$$

$$x^2 \approx (0.333)(2.6 \times 10^{-9}) = 8.7 \times 10^{-10}$$

$$x \approx 2.9 \times 10^{-5}$$

$$[OH^-] = x = 2.9 \times 10^{-5}$$

$$[H^+][OH^-] = K_w = 1.0 \times 10^{-14}$$

$$[H^+](2.9 \times 10^{-5}) = 1.0 \times 10^{-14}$$

$$[H^+] = \frac{1.0 \times 10^{-14}}{2.9 \times 10^{-5}} = 3.4 \times 10^{-10}\ M$$

$$pH = 9.47$$

**Test 5.13**

**A.** Since equal volumes are mixed, the concentrations are halved.

$$[Pb^{2+}]_0 = \frac{1.0 \times 10^{-3}}{2} = 5.0 \times 10^{-4}\ M$$

$$[Cl^-]_0 = \frac{1.0 \times 10^{-2}}{2} = 5.0 \times 10^{-3}\ M$$

For $PbCl_2(s)$, $Q = [Pb^{2+}]_0[Cl^-]_0^2$

$$= (5.0 \times 10^{-4})(5.0 \times 10^{-3})^2$$

$$= 1.25 \times 10^{-8}$$

$Q < K_{sp}$; no $PbCl_2(s)$ forms.

**B.** $[Pb^{2+}] = [Pb^{2+}]_0 = 5.0 \times 10^{-4}\ M$

$[Cl^-] = [Cl^-]_0 = 5.0 \times 10^{-3}\ M$

**Test 5.14**

**A.** $[Mg^{2+}]_0 = \dfrac{\text{mmol } Mg^{2+}}{\text{mL of solution}} = \dfrac{(150.0 \text{ mL})(1.00 \times 10^{-2} M)}{400.0 \text{ mL}} = 3.75 \times 10^{-3} M$

$[F^-]_0 = \dfrac{\text{mmol } F^-}{\text{mL of solution}} = \dfrac{(250.0 \text{ mL})(1.00 \times 10^{-1} M)}{400.0 \text{ mL}} = 6.25 \times 10^{-2} M$

$Q = [Mg^{2+}]_0[F^-]_0^2 = (3.75 \times 10^{-3})(6.25 \times 10^{-2})^2 = 1.46 \times 10^{-5}$

$Q > K_{sp}$; $MgF_2(s)$ will form

**B.** Run the precipitation reaction to completion.

| | $Mg^{2+}$ | + | $2F^-$ | $\longrightarrow$ | $MgF_2(s)$ |
|---|---|---|---|---|---|
| Before reaction | $(150.0)(1.00 \times 10^{-2})$ $= 1.5$ mmol | | $(250.0)(1.00 \times 10^{-1})$ $= 25.0$ mmol | | |
| After reaction | 0 | | $25.0 - 2(1.50)$ $= 22.0$ mmol | | |

**C.** Initial concentrations (after the reaction in part B has gone to completion):

$[Mg^{2+}]_0 = 0$

$[F^-]_0 = \dfrac{22.0 \text{ mmol}}{400.0 \text{ mL}} = 5.50 \times 10^{-2} M$

**D.** Let $x$ mol/L $MgF_2(s)$ dissolve to come to equilibrium.

$xMgF_2(s) \longrightarrow xMg^{2+} + 2xF^-$

**E.** Equilibrium concentrations:

$[Mg^{2+}] = x$

$[F^-] = 5.50 \times 10^{-2} M + 2x$

**F.** $6.4 \times 10^{-9} = K_{sp} = [Mg^{2+}][F^-]^2 = (x)(5.50 \times 10^{-2} + 2x)^2$

Assume $5.50 \times 10^{-2} \gg 2x$, so

$6.4 \times 10^{-9} \approx (x)(5.50 \times 10^{-2})^2$

$x \approx \dfrac{6.4 \times 10^{-9}}{(5.50 \times 10^{-2})^2} = 2.1 \times 10^{-6} M$

The assumption is valid.

**G.** $[Mg^{2+}] = x = 2.1 \times 10^{-6} M$

$[F^-] = 5.50 \times 10^{-2} M + 2x = 5.50 \times 10^{-2} M$

**Test 5.15**

**A.** $[Ag^+]_0 = \dfrac{(150.0 \text{ mL})(1.0 \times 10^{-3} M)}{350.0 \text{ mL}} = 4.29 \times 10^{-4} M$

$[S_2O_3^{2-}]_0 = \dfrac{(200.0 \text{ mL})(5.0 M)}{350.0 \text{ mL}} = 2.86 M$

**B.** Since $[S_2O_3^{2-}]_0 \gg [Ag^+]_0$ and since $K_1$ and $K_2$ are both large, the dominant species will be $Ag(S_2O_3)_2^{3-}$. The net reaction will be

$$Ag^+ + 2S_2O_3^{2-} \longrightarrow Ag(S_2O_3)_2^{3-}$$

In this example,

$4.29 \times 10^{-4} M \text{ Ag} + 2(4.29 \times 10^{-4} M)S_2O_3^{2-} \longrightarrow 4.29 \times 10^{-4} M \text{ } Ag(S_2O_3)_2^{3-}$

$[Ag(S_2O_3)_2^{3-}] \approx 4.29 \times 10^{-4} M$

**C.** $[S_2O_3^{2-}] = [S_2O_3^{2-}]_{original} -$ amount consumed

$$= 2.86\ M - 2(4.29 \times 10^{-4}\ M) = 2.86\ M$$

**D.** Calculate $[Ag(S_2O_3)^-]$ from $K_2$:

$$3.9 \times 10^4 = K_2 = \frac{[Ag(S_2O_3)_2^{3-}]}{[Ag(S_2O_3)^-][S_2O_3^{2-}]} = \frac{4.29 \times 10^{-4}}{[Ag(S_2O_3)^-](2.86)}$$

$$[Ag(S_2O_3)^-] = \frac{4.29 \times 10^{-4}}{(3.9 \times 10^4)(2.86)} = 3.8 \times 10^{-9}\ M$$

Calculate $[Ag^+]$ from $K_1$:

$$7.4 \times 10^8 = K_1 = \frac{[Ag(S_2O_3)^-]}{[Ag^+][S_2O_3^{2-}]} = \frac{3.8 \times 10^{-9}}{[Ag^+](2.86)}$$

$$[Ag^+] = \frac{3.8 \times 10^{-9}}{(7.4 \times 10^8)(2.86)} = 1.8 \times 10^{-18}\ M$$

**E.** These results show that $[Ag(S_2O_3)_2^{3-}] \gg [Ag(S_2O_3)^-] \gg [Ag^+]$. The assumptions are valid.

### Test 5.16

$$[NH_3]_{original} = \frac{3.00\ M}{2} = 1.50\ M$$

$$[Cu^{2+}]_{original} = \frac{2.00 \times 10^{-3}\ M}{2} = 1.00 \times 10^{-3}\ M$$

Since $[NH_3]_{original} \gg [Cu^{2+}]_{original}$ and since $K_1$, $K_2$, $K_3$, and $K_4$ are all large, $Cu(NH_3)_4^{2+}$ will be the dominant species. The net reaction will be

$$Cu^{2+} + 4NH_3 \longrightarrow Cu(NH_3)_4^{2+}$$

In this case, $1.00 \times 10^{-3}\ M\ Cu^{2+}$ plus $4(1.00 \times 10^{-3}\ M)NH_3$ will produce $1.00 \times 10^{-3}\ M$ $Cu(NH_3)_4^{2+}$. At equilibrium,

$$[Cu(NH_3)_4^{2+}] \approx 1.00 \times 10^{-3}\ M$$

$$[NH_3] = [NH_3]_0 - NH_3 \text{ consumed} = 1.50\ M - 4(1.00 \times 10^{-3}\ M) = 1.50\ M$$

Calculate $[Cu(NH_3)_3^{2+}]$ from $K_4$:

$$1.55 \times 10^2 = K_4 = \frac{[Cu(NH_3)_4^{2+}]}{[Cu(NH_3)_3^{2+}][NH_3]} = \frac{1.00 \times 10^{-3}}{[Cu(NH_3)_3^{2+}](1.50)}$$

$$[Cu(NH_3)_3^{2+}] = \frac{1.00 \times 10^{-3}}{[Cu(NH_3)_3^{2+}](1.50)} = 4.30 \times 10^{-6}\ M$$

Calculate $[Cu(NH_3)_2^{2+}]$ from $K_3$:

$$1.00 \times 10^3 = K_3 = \frac{[Cu(NH_3)_3^{2+}]}{[Cu(NH_3)_2^{2+}][NH_3]} = \frac{4.30 \times 10^{-6}}{[Cu(NH_3)_2^{2+}](1.50)}$$

$$[Cu(NH_3)_3^{2+}] = \frac{4.30 \times 10^{-6}}{(1.00 \times 10^3)(1.50)} = 2.87 \times 10^{-9}\ M$$

Calculate $[Cu(NH_3)^{2+}]$ from $K_2$:

$$3.88 \times 10^3 = K_2 = \frac{[Cu(NH_3)_2^{2+}]}{[Cu(NH_3)^{2+}][NH_3]} = \frac{2.87 \times 10^{-9}}{[Cu(NH_3)^{2+}](1.50)}$$

$$[Cu(NH_3)^{2+}] = \frac{2.87 \times 10^{-9}}{(3.88 \times 10^3)(1.50)} = 4.93 \times 10^{-13}\ M$$

Calculate $[Cu^{2+}]$ from $K_1$:

$$1.86 \times 10^4 = \frac{[Cu(NH_3)^{2+}]}{[Cu^{2+}][NH_3]} = \frac{4.93 \times 10^{-13}}{[Cu^{2+}](1.50)}$$

$$[Cu^{2+}] = \frac{4.93 \times 10^{-13}}{(1.86 \times 10^4)(1.50)} = 1.77 \times 10^{-17}\ M$$

The assumptions are valid. $Cu(NH_3)_4^{2+}$ is clearly the dominant copper-containing component.

### Test 5.17

The reactions to be considered are

$$Ag_3PO_4(s) \rightleftarrows 3Ag^+ + PO_4^{3-} \qquad K_{sp} = 1.8 \times 10^{-18}$$

$$PO_4^{3-} + H_2O \rightleftarrows HPO_4^{2-} + OH^- \qquad K_b = \frac{K_w}{K_{a_3}} = \frac{1.00 \times 10^{-14}}{4.8 \times 10^{-13}} = 2.1 \times 10^{-2}$$

The reaction of $HPO_4^{2-}$ with water to form $H_2PO_4^-$ need not be considered, since it is a much weaker base than $PO_4^{3-}$.

The problem contains four unknowns. So far we have two equations (the two equilibrium expressions). More relationships are clearly needed. One relationship that is almost always useful is the material balance equation. Since both the $Ag^+$ and the $PO_4^{3-}$ come from $Ag_3PO_4(s)$, the following relationship holds:

$[Ag^+] = 3$ times concentration of phosphate in all forms

Since the assumption has been made that $H_2PO_4^-$ and $H_3PO_4$ are unimportant,

$$[Ag^+] = 3\{[PO_4^{3-}] + [HPO_4^{2-}]\}$$

When two terms are to be added, consider whether it is reasonable to suppose that one term is much larger than the other. What about the relative values of $[PO_4^{3-}]$ and $[HPO_4^{2-}]$ at equilibrium? Both species are involved in the $K_b$ expression:

$$K_b = \frac{K_w}{K_{a_3}} = 2.1 \times 10^{-2} = \frac{[OH^-][HPO_4^{2-}]}{[PO_4^{3-}]}$$

Note that the ratio of the concentrations of $HPO_4^{2-}$ and $PO_4^{3-}$ depends on the $[OH^-]$. Can we estimate a reasonable value of $[OH^-]$?

$[OH^-]$ will definitely be greater than $10^{-7}\ M$ ($PO_4^{3-}$ is a base). Suppose $[OH^-] = 10^{-5}\ M$. Then

$$\frac{[HPO_4^{2-}]}{[PO_4^{3-}]} = \frac{K_b}{[OH^-]} = \frac{2.1 \times 10^{-2}}{10^{-5}} = 2.1 \times 10^3 = 2100$$

In this case,

$$[HPO_4^{2-}] \gg [PO_4^{3-}]$$

On the other hand, if $[OH^-] = 10^{-2}\ M$,

$$\frac{[HPO_4^{2-}]}{[PO_4^{3-}]} = \frac{2.1 \times 10^{-2}}{10^{-2}} = 2.1$$

$$[HPO_4^{2-}] \approx 2[PO_4^{3-}]$$

So arbitrarily assume that the $[OH^-]$ at equilibrium will be small enough so that

$$[HPO_4^{2-}] \gg [PO_4^{3-}]$$

What this assumption really means is that essentially all of the $PO_4^{3-}$ produced when $Ag_3PO_4(s)$ dissolves will react with water to produce $HPO_4^{2-}$, so that the net reaction will be

$$Ag_3PO_4(s) + H_2O(l) \rightleftarrows 3Ag^+(aq) + HPO_4^{2-}(aq) + OH^-(aq)$$

which is the sum of the reactions

$$Ag_3PO_4(s) \rightleftarrows 3Ag^+ + PO_4^{3-} \qquad K_{sp} = 1.8 \times 10^{-18}$$
$$PO_4^{3-} + H_2O \rightleftarrows HPO_4^{2-} + OH^- \qquad K_b = 2.1 \times 10^{-2}$$

When reactions are summed, equilibrium constants are multiplied:

$$K = [Ag^+]^3[HPO_4^{2-}][OH^-]$$
$$= K_{sp}K_b$$
$$= (1.8 \times 10^{-18})(2.1 \times 10^{-2}) = 3.8 \times 10^{-20}$$

Now let $x$ = solubility of $Ag_3PO_4$ according to the above reaction: $x Ag_3PO_4$ reacts with $xH_2O$ to produce $3xAg^+$ plus $xHPO_4^{2-}$ plus $xOH^-$.

Thus, at equilibrium,

$$[Ag^+] = 3x$$
$$[HPO_4^{2-}] = [OH^-] = x$$
$$K = [Ag^+]^3[HPO_4^{2-}][OH^-] = (3x)^3(x)(x) = 3.8 \times 10^{-20}$$
$$27x^5 = 3.8 \times 10^{-20}$$
$$x^5 = \frac{3.8 \times 10^{-20}}{27} = 1.4 \times 10^{-21}$$
$$x = 6.7 \times 10^{-5} = \text{solubility}$$

Now check the assumption that $[HPO_4^{2-}] \gg [PO_4^{3-}]$:

$$K_b = \frac{[OH^-][HPO_4^{2-}]}{[PO_4^{3-}]} = 2.1 \times 10^{-2}$$

$$\frac{[HPO_4^{2-}]}{[PO_4^{3-}]} = \frac{K_b}{[OH^-]} = \frac{2.1 \times 10^{-2}}{6.7 \times 10^{-5}} = 3.1 \times 10^2 = 310$$

↖ from above

$$[HPO_4^{2-}] \gg [PO_4^{3-}]$$

The assumptions are correct.

The solubility of $Ag_3PO_4$ in water is $6.7 \times 10^{-5}$ mol/L. Compare this to the calculated value of $1.6 \times 10^{-5}$ mol/L when the basicity of $PO_4^{3-}$ is ignored.

### Test 5.18

**A.** $CuS(s) \rightleftarrows Cu^{2+} + S^{2-} \qquad K_{sp} = 8.5 \times 10^{-45}$

$$S^{2-} + H^+ \rightleftarrows HS^- \qquad K = \frac{1}{K_{a_2}} = 7.75 \times 10^{12}$$

$$HS^- + H^+ \rightleftarrows H_2S \qquad K = \frac{1}{K_{a_1}} = 9.80 \times 10^6$$

**B.** $S^{2-}$ and $HS^-$ are both effective bases (each reacts with $H^+$ to produce an equilibrium that lies far to the right—see part A). This means that essentially all of the $S^{2-}$ that is released into the solution as $CuS(s)$ dissolves will end up as $H_2S$. The conclusion is that $H_2S$ will be the dominant sulfur-containing species.

**C.** The net reaction is

$$CuS(s) + 2H^+(aq) \longrightarrow H_2S(aq) + Cu^{2+}(aq)$$

**D.** $K = \frac{[H_2S][Cu^{2+}]}{[H^+]^2} = \frac{K_{sp}}{K_{a_1} \cdot K_{a_2}} = 8.2 \times 10^{-25}$

**E.** $CuS(s) + 2H^+ \longrightarrow H_2S + Cu^{2+}$

$x$CuS($s$) reacts with $2xH^+$ to produce $xH_2S$ plus $xCu^{2+}$

At equilibrium: $[Cu^{2+}] = x$

$[H_2S] = x$

$[H^+] = 10.0 - \text{amount consumed} = 10.0 - 2x$

↖ initial concentration

$$6.5 \times 10^{-25} = \frac{[H_2S][Cu^{2+}]}{[H^+]^2} = \frac{(x)(x)}{(10.0 - 2x)^2}$$

Note that $K$ is small, which means that $2x$ will be small compared to 10.0. Thus

$$6.5 \times 10^{-25} = \frac{(x)(x)}{(10 - 2x)^2} \approx \frac{x^2}{(10)^2}$$

$$x^2 \approx 6.5 \times 10^{-23}$$

$$x = 8.1 \times 10^{-12} \text{ mol/L}$$

Check the assumption that $H_2S$ is dominant:

$$[H^+] = 10.0 - 2x = 10.0 - 2(8.1 \times 10^{-12}) = 10.0\ M$$

Using $K_{a_1}$:

$$K_{a_1} = 1.02 \times 10^{-7} = \frac{[H^+][HS^-]}{[H_2S]}$$

$$\frac{[H_2S]}{[HS^-]} = \frac{[H^+]}{K_{a_1}} = \frac{(10.0)}{(1.02 \times 10^{-7})} = 9.8 \times 10^7$$

$$[H_2S] \gg [HS^-]$$

Using $K_{a_2}$:

$$K_{a_2} = 1.29 \times 10^{-13} = \frac{[H^+][S^{2-}]}{[HS^-]}$$

$$\frac{[HS^-]}{[S^{2-}]} = \frac{[H^+]}{K_{a_2}} = \frac{(10.0)}{(1.29 \times 10^{-13})} = 7.8 \times 10^{13}$$

$$[HS^-] \gg [S^{2-}]$$

In summary: $[H_2S] \gg [HS^-] \gg [S^{2-}]$. $H_2S$ is dominant. The solubility of CuS in 10.0 $M$ $H^+$ is $8.1 \times 10^{-12}$ mol/L.

### Test 5.19

**A.** $AgI(s) \rightleftarrows Ag^+ + I^-$ $\qquad K_{sp} = 1.5 \times 10^{-16}$

$Ag^+ + S_2O_3^{2-} \rightleftarrows Ag(S_2O_3)^-$ $\qquad K_1 = 7.4 \times 10^8$

$AgS_2O_3^- + S_2O_3^{2-} \rightleftarrows Ag(S_2O_3)_2^{3-}$ $\qquad K_2 = 3.9 \times 10^4$

**B.** $Ag(S_2O_3)_2^{3-}$ will be dominant.

**C.** $AgI(s) + 2S_2O_3^{2-} \rightleftarrows Ag(S_2O_3)_2^{3-} + I^-$

**D.** $K = \frac{[Ag(S_2O_3)_2^{3-}][I^-]}{[S_2O_3^{2-}]^2} = K_{sp}K_1K_2 = 4.3 \times 10^{-3}$

**E.** Let $x$ = solubility of AgI($s$) at equilibrium:

$$[I^-] = [Ag(S_2O_3)_2^{3-}] = x$$

$$[S_2O_3^{2-}] = 5.0\ M - 2x$$

$$4.3 \times 10^{-3} = K = \frac{[Ag(S_2O_3)_2^{3-}][I^-]}{[S_2O_3^{2-}]^2} = \frac{(x)(x)}{(5.0 - 2x)^2}$$

Assume $5.0 - 2x \approx 5.0$, so

$$4.3 \times 10^{-3} \approx \frac{x^2}{(5.0)^2}$$

$$x^2 \approx (5.0)^2(4.3 \times 10^{-3}) = 1.1 \times 10^{-1}$$

$$x \approx 3.3 \times 10^{-1}$$

Check the assumption:

$$2x = 2(0.33) = 0.66$$

$$\frac{0.66}{5.0} \times 100 = 13.2\%$$

The approximation is not valid. The expression

$$\frac{x^2}{(5.0 - 2x)^2} = 4.3 \times 10^{-3}$$

must be solved directly (take the square root of both sides). This gives

$$x = 0.29\ \text{mol/L}$$

$$[S_2O_3^{2-}] = 5.0 - 2x = 5.0 - 0.58 = 4.4\ M$$

Now check the original assumption that $[Ag(S_2O_3)_2^{3-}]$ is dominant.
Using $K_2$:

$$K_2 = \frac{[Ag(S_2O_3)_2^{3-}]}{[Ag(S_2O_3)^-][S_2O_3^{2-}]} = 3.9 \times 10^4$$

$$\frac{[Ag(S_2O_3)_2^{3-}]}{[Ag(S_2O_3)^-]} = K_2[S_2O_3^{2-}] = (3.9 \times 10^4)(4.4) = 1.7 \times 10^5$$

$$[Ag(S_2O_3)_2^{3-}] \gg [Ag(S_2O_3)^-]$$

Using $K_1$:

$$K_1 = \frac{[Ag(S_2O_3)^-]}{[Ag^+][S_2O_3^{2-}]} = 7.4 \times 10^8$$

$$\frac{[Ag(S_2O_3)^-]}{[Ag^+]} = K_1[S_2O_3^{2-}] = (7.4 \times 10^8)(4.4) = 3.3 \times 10^9$$

$$[Ag(S_2O_3)^-] \gg [Ag^+]$$

In summary: $[Ag(S_2O_3)_2^{3-}] \gg [Ag(S_2O_3)^-] \gg [Ag^+]$. The solubility of AgI(*s*) in 5.0 *M* $Na_2S_2O_3$ is 0.29 mol/L.

# Answers to Exercises

## Chapter 1

1. (a) $K = \dfrac{[H_2O]^2}{[H_2]^2[O_2]}$
   (b) $K = \dfrac{[H_2O]^2}{[H_2]^2}$
   (c) $K = \dfrac{[NO_2]^4[H_2O]^6}{[NH_3]^4[O_2]^7}$
2. $K = \dfrac{[PCl_5]}{[PCl_3][Cl_2]}$
3. (a) 11 L/mol (b) $9.0 \times 10^{-2}$ mol/L
4. left
5. right
6. $[SO_2] = 0.34$ mol/L
7. (a) $1.7 \times 10^{-3}$ atm$^{-2}$ (b) smaller
8. (a) 45.2 (b) 45.2
9. $[HI] = 4.30$ mol/L $[I_2] = 4.10$ mol/L
10. $1.0 \times 10^1$ L/mol
11. (a) $[CO_2] = 8.0 \times 10^{-3}$ mol/L (b) $4.0 \times 10^{-2}$
    (c) $4.0 \times 10^{-2}$ (d) $6.7 \times 10^{-1}$ atm
12. $1.7 \times 10^{12}$ L/mol
13. $6.9 \times 10^{10}$ atm$^{-1}$
14. (a) shifts left (b) shifts left
    (c) no change (d) shifts right
15. $2.4 \times 10^{-2}$ mol/L
16. (a) no. $Q > K$
    (b) will shift left
    (c) $2x$ = mol/L of $NH_3$ formed to reach equilibrium (other choices are possible)
    (d) $[NH_3] = 2.0 \times 10^{-4} + 2x$; $[N_2] = 2.0 \times 10^{-1} - x$; $[H_2] = 2.0 \times 10^{-2} - 3x$
17. $K_p = \dfrac{P_{NO_2}{}^2}{P_{N_2O_4}} = \dfrac{(6.10 \times 10^{-2})^2}{1.33 \times 10^{-2}} = 2.80 \times 10^{-1}$ atm

54. $[PCl_5] = 1.8 \times 10^{-3}$ mol/L
    $[PCl_3] = 6.5 \times 10^{-2}$ mol/L
    $[Cl_2] = 2.5 \times 10^{-3}$ mol/L
55. $[NO_2] = 7.2 \times 10^{-2}$ mol/L
    $[N_2O_4] = 6.4 \times 10^{-2}$ mol/L
56. $[NH_3] = 4.6 \times 10^{-3}$ atm
    $[N_2] = 1.0 \times 10^{-1}$ atm
    $[H_2] = 4.9 \times 10^{-1}$ atm
60. $[Cl_2] = 9.6 \times 10^{-2}$ mol/L
    $[PCl_5] = 1.84 \times 10^{-1}$ mol/L
    $[PCl_3] = 4.6 \times 10^{-2}$ mol/L
61. $[HI] = 6.90 \times 10^{-1}$ mol/L
    $[H_2] = 1.55 \times 10^{-1}$ mol/L
    $[I_2] = 5.0 \times 10^{-3}$ mol/L
57. (a) left
    (b) $[NH_3] = 1.0 \times 10^{-3}$ atm
    $[N_2] = 1.3 \times 10^{-1}$ atm
    $[H_2] = 1.7 \times 10^{-1}$ atm
58. $[HI] = 2.35$ mol/L
    $[H_2] = [I_2] = 0.377$ mol/L
59. (a) shifts right
    (b) $[HI] = 2.62$ mol/L
    $[H_2] = 0.241$ mol/L
    $[I_2] = 0.741$ mol/L
62. (a) no; $Q > K$
    (b) will shift left
    (c) $[ClO] = 3.0 \times 10^{-2}$ mol/L
    $[Cl_2] = 2.3 \times 10^{-5}$ mol/L
    $[O_2] = 2.5 \times 10^{-1}$ mol/L

## Chapter 2

1. (a) $HNO_2$, $H_2O$
   $HNO_2 \rightleftarrows H^+ + NO_2^-$
   (b) $H^+$, $NO_3^-$, $H_2O$
   $H^+$ from complete dissociation of $HNO_3$
   (c) $C_5H_5N$, $H_2O$
   $C_5H_5N + H_2O \rightleftarrows C_5H_5NH^+ + OH^-$
   (d) $Na^+$, $CN^-$, $H_2O$
   $CN^- + H_2O \rightleftarrows HCN + OH^-$

(e) $HCHO_2$, $Na^+$, $CHO_2^-$, $H_2O$
$HCHO_2 \rightleftarrows H^+ + CHO_2^-$ or
$CHO_2^- + H_2O \rightleftarrows HCHO_2 + OH^-$
(f) $Na^+$, $OH^-$, $H_2O$
$OH^-$ from complete dissociation of NaOH
(g) $Pu^{3+}(aq)$, $NO_3^-$, $H_2O$
$Pu(OH_2)_6^{3+} \rightleftarrows Pu(OH)(OH_2)_5^{2+} + H^+$
(h) HOCl, $H_2O$
$HOCl \rightleftarrows H^+ + OCl^-$
(i) $(CH_3)_3N$, $H_2O$
$(CH_3)_3N + H_2O \rightleftarrows (CH_3)_3NH^+ + OH^-$
(j) $H^+$, $Cl^-$, $H_2O$
$H^+$ from dissolved HCl
(k) $K^+$, $OH^-$, $H_2O$
$OH^-$ from dissolved KOH
(l) $HNO_2$, $Na^+$, $NO_2^-$, $H_2O$
$HNO_2 \rightleftarrows H^+ + NO_2^-$
(m) HF, $Na^+$, $F^-$, $H_2O$
$HF \rightleftarrows H^+ + F^-$
(n) $CH_3NH_2$, $H_2O$
$CH_3NH_2 + H_2O \rightleftarrows CH_3NH_3^+ + OH^-$
(o) $Al^{3+}(aq)$, $NO_3^-$, $H_2O$
Assume $Al^{3+}(aq)$ is $Al(OH_2)_6^{3+}$ since the $K_a$ for $Al(OH_2)_6^{3+}$ is given. $Al(OH_2)_6^{3+} \rightleftarrows Al(OH)(OH_2)_5^{2+} + H^+$

2. (a) 1.70
(b) 2.00
(c) 9.57
(d) 11.10
(e) 3.44
(f) 14.00
(g) 4.48
(h) 3.73
(i) 11.86
(j) 1.00
(k) 11.00
(l) 3.10
(m) 3.54
(n) 12.17
(o) 3.43

3. b

4. $K_a = 1.0 \times 10^{-7}$

5. (a) 2.49
(b) 2.11
(c) 2.63

6. 8.44

7. 9.26

8. (a) 4.98
(b) 4.70

---

## Chapter 3

1. (a) (1) $H^+$, $Cl^-$, $Na^+$, $OH^-$, $H_2O$
(2) $H^+$, $+ OH^- \longrightarrow H_2O$
(3) $Na^+$, $OH^-$, $H_2O$
(4) Excess $OH^-$ will determine pH
(b) (1) $Na^+$, $OH^-$, HOAc, $H_2O$
(2) $OH^- + HOAc \longrightarrow H_2O + OAc^-$
(3) HOAc, $OAc^-$, $Na^+$, $H_2O$
(4) $HOAc \rightleftarrows H^+ + OAc^-$ or
$OAc^- + H_2O \rightleftarrows HOAc + OH^-$
(c) (1) $H^+$, $Cl^-$, $NH_3$, $H_2O$
(2) $NH_3 + H^+ \longrightarrow NH_4^+$
(3) $NH_3$, $NH_4^+$, $Cl^-$, $H_2O$
(4) $NH_4^+ \rightleftarrows NH_3 + H^+$ or
$NH_3 + H_2O \rightleftarrows NH_4^+ + OH^-$
(d) (1) $H^+$, $Cl^-$, $Na^+$, $CN^-$, $H_2O$
(2) $H^+ + CN^- \longrightarrow HCN$
(3) HCN, $CN^-$, $Na^+$, $Cl^-$, $H_2O$
(4) $HCN \rightleftarrows H^+ + CN^-$ or
$CN^- + H_2O \rightleftarrows HCN + OH^-$
(e) (1) $Na^+$, $OH^-$, $HCHO_2$, $H_2O$
(2) $OH^- + HCHO_2 \longrightarrow H_2O + CHO_2^-$
(3) $HCHO_2$, $CHO_2^-$, $Na^+$, $H_2O$
(4) $HCHO_2 \rightleftarrows H^+ + CHO_2^-$ or
$CHO_2^- + H_2O \rightleftarrows HCHO_2 + OH^-$
(f) (1) $C_6H_5COOH$, $Na^+$, $OH^-$, $H_2O$
(2) $C_6H_5COOH + OH^- \longrightarrow C_6H_5COO^- + H_2O$
(3) $C_6H_5COO^-$, $Na^+$, $OH^-$, $H_2O$
(4) excess $OH^-$ from NaOH
(g) (1) $H^+$, $NO_3^-$, $C_5H_5N$, $H_2O$
(2) $H^+ + C_5H_5N \longrightarrow C_5H_5NH^+$
(3) $C_5H_5N$, $C_5H_5NH^+$, $NO_3^-$, $H_2O$
(4) $C_5H_5NH^+ \rightleftarrows C_5H_5N + H^+$ or
$C_5H_5N + H_2O \rightleftarrows C_5H_5NH^+ + OH^-$
(h) (1) $NH_3$, $H^+$, $Cl^-$, $H_2O$
(2) $NH_3 + H^+ \longrightarrow NH_4^+$
(3) $NH_3$, $NH_4^+$, $Cl^-$, $H_2O$
(4) $NH_4^+ \rightleftarrows NH_3 + H^+$ or
$NH_3 + H_2O \rightleftarrows NH_4^+ + OH^-$

2. a. 12.70
b. 4.60
c. 9.86
d. 9.21
e. 4.22
f. 13.65
g. 6.10
h. 9.26

3. a. 0.70
b. 1.10
c. 1.30
d. 7.00
e. 12.60

4. a. 13.30
b. 13.11
c. 12.88
d. 12.52

e. 7.00
f. 1.65

5. a. (1) 4.95
(2) 8.91
(3) 9.21
(4) 11.02
(5) 12.60
b. $K_a \approx 10^{-12}$

6. a. (1) yellow
(2) 3.30
(3) yellow to yellow-orange
b. (1) red
(2) 5.30
(3) red to red-orange
c. 4.3

7. a. 11.28
b. 10.21
c. 9.73
d. 9.26
e. 5.33
f. 1.65

## Chapter 4

1. $5.3 \times 10^{-12}$
2. $6.9 \times 10^{-9}$
3. $3.9 \times 10^{-5}$ mol/L
4. $8.9 \times 10^{-4}$ mol/L
5. $1.1 \times 10^{-29}$
6. (a) $8.8 \times 10^{-7}$ mol/L
(b) $7.7 \times 10^{-11}$ mol/L
7. $1.0 \times 10^{-10}$
8. (a) $1.6 \times 10^{-9}$ mol/L
(b) $1.6 \times 10^{-20}$ mol/L
9. $5.2 \times 10^{-9}$ mol/L
10. (a) $CaF_2(s)$
(b) $FePO_4(s)$
11. $AgCN(s)$ will form
12. $Ce(IO_3)_3(s)$ will form
13. $[S^{2-}]$ greater than $7.0 \times 10^{-15}$ mol/L
14. $CaSO_4(s)$ will not form
15. $CuCl(s)$ will form
16. $PbF_2(s)$ will not form
17. $[Ag^+]$ greater than $5.6 \times 10^{-5}$ mol/L
18. $2.6 \times 10^{-13}$

## Chapter 5

1. 2.51
2. 6.89
3. 2.28
4. 6.69
5. 0.65
6. (a) 1.70 (b) 4.69
(c) 2.30 (d) 11.11
7. 0.04
8. (a) 1.36 (b) 2.03
9. 2.32
10. $[Ce^{3+}] = 8.2 \times 10^{-6}\ M$
$[IO_3^-] = 3.5 \times 10^{-2}\ M$
11. $[BeF_4^{2-}] = 5.0 \times 10^{-5}\ M$
$[F^-] = 4.0\ M$
$[BeF_3^-] = 4.6 \times 10^{-7}\ M$
$[BeF_2] = 1.9 \times 10^{-10}\ M$
$[BeF^+] = 8.2 \times 10^{-15}\ M$
$[Be^{2+}] = 2.6 \times 10^{-20}\ M$
12. $6.0 \times 10^{-2}$ mol/L
13. $8.7 \times 10^{-16}$ mol/L
14. $1.3 \times 10^{-1}$ mol/L
(The answer is $1.27 \times 10^{-1}$ mol/L, assuming one extra significant figure.)
15. $2.5 \times 10^{-4}$ mol/L
16. $2.0 \times 10^{-6}$ mol/L

## Chapter 6

1. (a) 

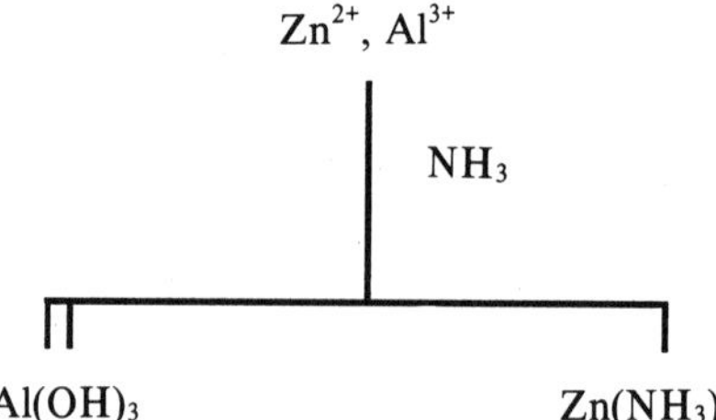


(b)

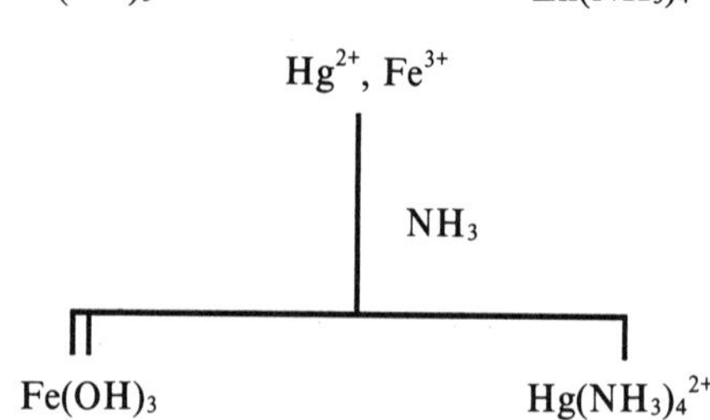


(c)

$Ba^{2+}$, $Pb^{2+}$

cold HCl

$PbCl_2$ $Ba^{2+}$

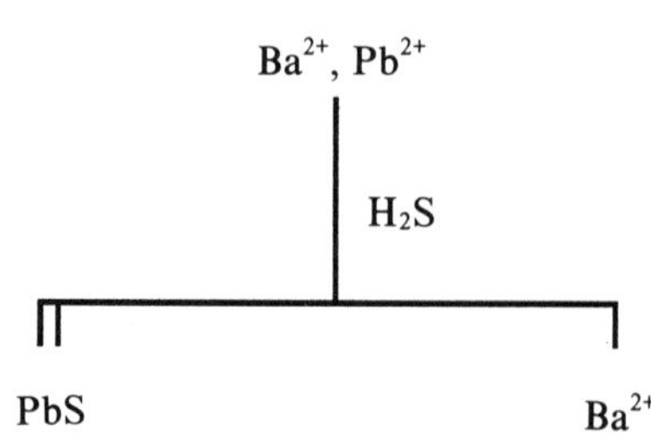


(d)

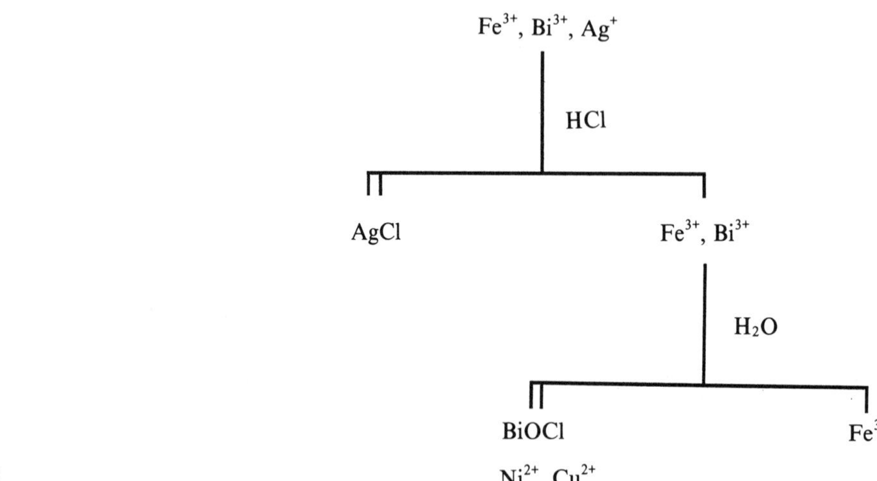


(e)

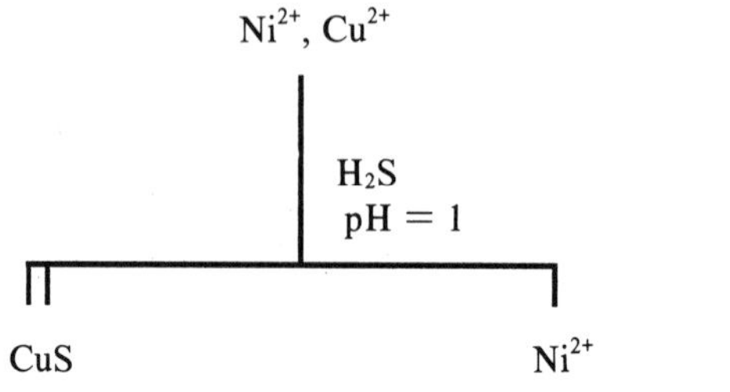


2. (a) $Ba^{2+}(aq) + CrO_4^{2-}(aq) \longrightarrow BaCrO_4(s)$
   (b) $Hg^{2+}(aq) + 4NH_3(aq) \longrightarrow Hg(NH_3)_4^{2+}(aq)$
   (c) See section 6.8
   (d) $3Ni^{2+}(aq) + 2PO_4^{3-}(aq) \longrightarrow Ni_3(PO_4)_2(s)$

3. (a) The tan precipitate is sulfur formed by oxidation of $S^{2-}$ by $Fe^{3+}$.
   (b) $Fe(OH)_3$ precipitates
   (c) The tan precipitate is sulfur [see (a)]. The black precipitate is FeS ($Fe^{2+}$ is produced by reduction of $Fe^{3+}$ by $S^{2-}$).

# Answers to Multiple-Choice Questions

**Chapter 1**

24. c
25. a
26. a
27. a
28. e
29. b
30. c
31. c
32. c
33. b
34. c
35. (1) c
    (2) c
    (3) b
    (4) b
36. c
37. b
38. b
39. a
40. a
41. b
42. d
43. c
44. b
45. b
46. c
47. c
48. a
48. c
50. d
51. b
52. b
53. c

**Chapter 2**

16. c
17. c
18. d
19. b
20. a
21. c
22. d
23. (a) c
    (b) d
24. c
25. a
26. a
27. a
28. b
29. a
30. e[1]
31. c
32. d
33. b
34. c
35. b
36. c
37. d
38. a
39. b
40. d[2]
41. c
42. c
43. c
44. b
45. a
46. c
47. a
48. c
49. c
50. b
51. d
52. c

**Chapter 3**

13. c
14. a
15. a
16. b
17. d
18. e
19. b
20. b
21. d
22. c
23. c
24. c
25. d
26. b
27. c
28. c
29. d
30. b
31. d
32. b
33. d
34. a
35. c
36. a
37. c
38. d
39. a
40. c
41. a
42. d
43. a
44. b[3]
45. a
46. b
47. d
48. d
49. b
50. d
51. a
52. c
53. d
54. b

**Chapter 4**

23. a
24. e
25. a
26. a
27. a
28. a
29. c
30. d
31. d
32. b
33. d
34. b
35. a

**Chapter 5**

17. d
18. d
19. a
20. e
21. c
22. c
23. d
24. d
25. a
26. d
27. d
28. a

[1] answer is $1.95 \times 10^{-6}$ $M$

[2] answer is $2.0 \times 10^{-4}$ $M$

[3] not using quadratic formula